Energy, Power and Protest on the Urban Grid

Providing a global overview of experiments around the transformation of cities' electricity networks and the social struggles associated with this change, this book explores the centrality of electricity infrastructures in the urban configuration of social control, segregation, integration, resource access and poverty alleviation. Through multiple accounts from a range of global cities, this edited collection establishes an agenda that recognises the uneven, and often historical, geographies of urban electricity networks, prompting attempts to re-wire the infrastructure configurations of cities and predicating protest and resistance from residents and social movements alike. Through a robust theoretical engagement with established work around the politics of urban infrastructures, the book frames the transformation of electricity systems in the context of power and resistance across urban life, drawing links between environmental and social forms of sustainability. Such an agenda can provide both insight and inspiration in seeking to build fairer and more sustainable urban futures that bring electricity infrastructures to the fore of academic and policy attention.

Andrés Luque-Ayala is a Lecturer at Durham University's Geography Department. His research examines the emergence of a local governance of energy and the interface between urban infrastructures, climate change and development modes in the global South.

Jonathan Silver is a Leverhulme Early Career Research Fellow at Durham University's Geography Department. His research works at the intersection of urban infrastructure, politics and socio-environmental inequalities.

Energy, Power and Protest on the Urban Grid

Geographies of the Electric City

Edited by
**Andrés Luque-Ayala and
Jonathan Silver**

Routledge
Taylor & Francis Group

LONDON AND NEW YORK

First published 2016
by Routledge
2 Park Square, Milton Park, Abingdon, Oxon OX14 4RN

and by Routledge
711 Third Avenue, New York, NY 10017

Routledge is an imprint of the Taylor & Francis Group, an informa business

© 2016 selection and editorial matter, Andrés Luque-Ayala and Jonathan Silver; individual chapters, the contributors

British Library Cataloguing in Publication Data
A catalogue record for this book is available from the British Library

Library of Congress Cataloging in Publication Data
Names: Luque-Ayala, Andrés, editor.
Title: Energy, power and protest on the urban grid: geographies of the
    electric city / [edited] by Andrés Luque-Ayala and Jonathan Silver.
Description: Farnham, Surrey, UK; Burlington, VT: Ashgate, [2016] |
    Includes bibliographical references and index.
Identifiers: LCCN 2015034301| ISBN 9781472449009 (hardback: alk.
    paper) | ISBN 9781472449016 (ebook) | ISBN 9781472449023 (epub)
Subjects: LCSH: Electric utilities—Political aspects. | Energy
    consumption—Political aspects. | Discrimination. | Social movements.
Classification: LCC HD9685.A2 E5757 2016 | DDC 333.793/2—dc23

ISBN: 9781472449009 (hbk)
ISBN: 9781315579597 (ebk)

Typeset in Times New Roman
by Apex CoVantage, LLC

MIX
Paper from
responsible sources
FSC® C013985
www.fsc.org

Printed in the United Kingdom
by Henry Ling Limited

Contents

Figures

About the Authors

Editors

Andrés Luque-Ayala is a lecturer at Durham University's Geography Department. His research focuses on the socio-political dimensions of emerging infrastructural narratives and configurations in cities in the global South, including the emergence of a local governance of energy and the interface between digital technologies, ecological security and development modes. He holds a PhD in Geography (Durham), as well as an MSc in City Design (LSE), a Masters in Environmental Management (Yale) and first degrees in Anthropology and Political Science (Universidad de los Andes, Colombia).

Jonathan Silver is a research fellow at Durham University's Geography Department. His research works at the intersection of urban infrastructure, politics and socio-environmental inequality. He holds a PhD in Geography (Durham), together with an MA in Geography (University of Manchester).

Chapter Contributors

Georgia Alexandri is a postdoctoral researcher at the Autonomous University of Madrid (UAM). She holds a PhD in Human Geography from the Department of Geography of Harokopio University (Athens) and an MSc in Sustainability and Planning (Cardiff University). Her research interests focus on gentrification, socio-spatial restructuring, financialisation of housing markets, fear of the 'other' and social movements.

Rosalina Babourkova is a researcher with the Urban Management Program at Berlin Institute of Technology. She holds a PhD in Planning Studies and a Masters in Environment and Sustainable Development (University College London). Her research experience lies in urban infrastructures, space and subjectivities, while her current research interests relate to the imbrication of urban space, motherhood and transgression.

Idalina Baptista is Associate Professor in Urban Anthropology at the University of Oxford. Her current research focuses on the colonial and post-colonial geographies of urban energy infrastructure and urbanisation in African cities, with a special focus on Maputo, Mozambique. She holds a PhD in City and Regional Planning (UC Berkeley), a Masters in Landscape Architecture (UC Berkeley) and a BEng in Environmental Engineering (Universidade Nova de Lisboa, Portugal).

Venetia Chatzi is a PhD student of human geography at Harokopio University in Athens, Greece. Her research interests centre around urban governance, management and public space use. She has an MSc in Human Geography (Harokopio University) and a BA in International Relations and Politics (University of Macedonia in Thessaloniki). Venetia works as an executive administrator for the Green Institute in Athens and collaborates with NGOs on projects working towards the protection of the environment and promotion of intercultural dialogue.

Arwen Colell is a guest lecturer at CIEE Berlin. She studied politics in Germany, Japan and the US. Her PhD dissertation at the Environmental Policy Research Center in Berlin, supported by a Heinrich Böll Foundation scholarship, focuses on community energy projects. In the past Arwen worked for the Berlin-based energy start-up Ubitricity – where she focused on the smart grid integration of electric vehicles – and for the German Federal Ministry of the Environment.

Laure Criqui is a research fellow for international urban development at the think-tank IDDRI in Paris. Her research interests focus on the extension of basic services in irregular settlements, the innovations utilities use for this purpose and their potential to renew both practices and understandings of urban planning. She has practitioner experience in international cooperation projects, and holds a PhD in Urban Planning (LATTS/Université Paris-Est) and an MSc in Geography (LSE).

Conor Harrison is an assistant professor at the Department of Geography, Environment and Sustainability Program, University of South Carolina. His research examines how American society has shaped and continues to shape its energy infrastructures, and conversely, how energy infrastructures shape the American consumption of energy. He holds a PhD and an MA in Geography (University of North Carolina and East Carolina University, respectively) and a BA in Political Science (Colgate University).

Anne Maassen is the Energy, Climate, and Finance Associate for the Energy & Climate practice area at the WRI Ross Center for Sustainable Cities, Washington D.C. She holds a PhD in Geography (Durham), as well as an MSc in Environmental Monitoring, Modelling and Management (King's College London).

Luise Neumann-Cosel is co-founder and manager of BürgerEnergie Berlin eG, a citizens' cooperative aiming to buy the local electricity grid. She studied geo-ecology in Germany and Panama, and has worked as a research associate in energy politics for the Berlin parliament as well as for several NGOs. Luise is an activist working on energy issues in Germany with experience in advancing Germany's anti-nuclear campaigns.

Francesca Pilo' is a postdoctoral researcher in urban studies at CESSMA, University Paris Diderot. She works in the DALVAA programme 'Rethinking the right to the city from the cities of the South'. Francesca holds a PhD in Urban Planning (University of Paris-Est and Fluminense Federal University, Brazil). Her research focuses the urban experiences of city dwellers in informal settlements, focusing specifically on electricity access, spatial inequalities in service distribution, and the meaning of the right to the city through socio-technical systems.

Eric Verdeil is a research fellow at the French National Centre for Scientific Research. He recently joined LATTS at Paris-Est University, after working with the Environnement Ville Société research laboratory at the University of Lyon. He teaches on the Master's programme Governing the Large Metropolis at Sciences Po, Paris. His research focuses on urban infrastructure, city planning and governance. Geographically, he focuses on Middle Eastern and North African cities, particularly Tunis and Sfax in Tunisia, Beirut in Lebanon, and Amman in Jordan.

Acknowledgements

As editors it has been a pleasure to work with the contributors and we would like to extend our thanks for their important role in the development of this book. We would also like to thank the team at Routledge and Ashgate, particularly Katy Crossan, for initial encouragement and ongoing support.

The book has benefited significantly from our work together at Durham University's Geography Department and at the Durham Energy Institute (DEI), in the United Kingdom. At both institutions we were given time, space and intellectual support to think through the issues that helped to bring the publication together. The DEI's Centre for Doctoral Training in Energy also provided us with financial assistance towards the editing process. Several of the ideas that form this book were also developed in the context of workshops and discussion sessions organised with the support of the Energy Geographies Working Group of the Royal Geographical Society (RGS-IBG). We would like to extend our sincere thanks to these institutions and to the people behind them.

We would also like to thank Harriet Bulkeley, Cheryl McEwan and Colin McFarlane who helped us develop our energy-related ideas. Finally, a very special thank you to Simon Marvin who supported us unconditionally through our postdoctoral work.

Jonathan would also like to acknowledge the ESRC grant entitled "Urban Transitions: Climate Change, Global Cities, and the Transformation of Socio-Technical Networks" [Award number RES-066–27–0002] and the EPSRC/DFID/DECC-funded project entitled "SAMSET" [Grant Number EP/L002620/1].

The Editors

Foreword

Consider the electric light by which you may be reading these words. What might it mean to think of this mundane, although far from pervasive, infrastructure of illumination as giving shape to contemporary urban politics? In what ways are access to electricity, and how and by whom electricity services are provided, now an important political terrain?

At a time in which energy is widely acknowledged to be one of the 'grand challenges' facing societies around the world, there is growing recognition among researchers, activists and the urban policy community of the ways in which energy infrastructures give structure and order to social life. Infrastructure slipped its established epistemological moorings within engineering some time ago: the wires, devices and networks that make up the physical geographies of energy systems are now a legitimate object of social science, and increasingly examined for the political work they perform. Urban electrical networks not only ferry electrons: their geographical reach, terms of access and forms of ownership reflect prevailing distributions of economic and political power. At the same time, electricity's topologies of connection and isolation, and the social experiences and aspirations it enables, create new publics, political identities and spaces for resistance. Observations like these, and many others that spring from the pages of this book, open up space for a critical urban energy politics. Such an approach embeds energy technologies large and small – from photovoltaic arrays and transmission networks to distribution grids and pre-payment meters – within power-laden social structures, and refuses to separate apparently technological and managerial issues (about, say, the design and operation of a urban-scale distribution network) from a consideration of their political potential.

A critical interest in the politics of energy infrastructure has arguably never been more important. Addressing the multiple dimensions of today's 'energy challenge' will require major interventions in the physical infrastructures through which energy is captured, transformed, distributed and consumed. The IEA (2014), for example, estimates an eye-watering $53 trillion is needed worldwide in cumulative infrastructural investment by 2035, to secure a reliable energy system consistent with limiting global temperature rise to 2°C. By comparison, the IEA have estimated the investment required to deliver universal access to modern energy services at around $750 billion (2010). Regardless of whether the policy priority is to address energy poverty, achieve universal access to modern energy services or mitigate climate change, global energy futures will be shaped by the rate and form of investment in infrastructure over the next few years. How this infrastructure will be owned and run makes a difference: for example, the re-municipalisation of urban electrical supply contracts in the last few years, in the wake of two decades of privatisation, reflects growing municipal concerns about loss of control over a critical component of urban planning as well as a desire among residents for more participatory forms of economic democracy (Cumbers, 2012). Whether managed as a financial asset, wielded as a developmental tool of the local state or imagined as a new urban commons, infrastructural investments that 'rewire' the city have the capacity to create new geographies of uneven development and conditions of political possibility. They are potentially as profound as those associated with coal-fired industrialisation or the advent of urban electrical light and power towards the end of the nineteenth century.

Energy, Power and Protest on the Urban Grid is a richly empirical collection of chapters that highlights what is at stake as urban electrical systems around the world are reworked in response to fiscal, democratic and environmental demands. Electrical networks have already proven to be very good for thinking with: Thomas Hughes' (1983) seminal work on the emergence of large regional power systems in the United States; Harold Platt's (1991) study of the electrification of Chicago; David McDonald's (2009) reflections on 'electric capitalism' in southern Africa; and Jane Bennett's (2005) essay on the North American blackout illustrate some of the different ways in which the 'politics' of networked electrical infrastructure have previously been interpreted. Contributors to this collection adopt an explicitly socio-technical approach to urban infrastructure, building on a decade and more of research on the politics of water, sanitation and other systems necessary for the reproduction of urban life. Chapters affirm that the 'modern infrastructural ideal' – universal access guided by an ethic of public service – has been highly specific, historically and geographically, in relation to electricity provision. 'Splintering urbanism' provides an entry point for many of the contributors (Graham & Marvin, 2001) to make sense of the variegated and partial implementation of this ideal, and for understanding how technical, administrative and commercial interventions in electricity infrastructure create political and economic differentiation at the urban scale. Other chapters draw inspiration from Tim Mitchell's (2011) examination of the different political possibilities afforded by coal and oil infrastructures in the twentieth century, to explore how electricity networks and their constituent components distribute political agency and nurture particular identities (*vis-à-vis* the state, for example).

A significant strength of *Energy, Power and Protest on the Urban Grid* is the diversity of its urban case material, encompassing small and medium-size cities (Plovdiv in Bulgaria, Rocky Mount in North Carolina), urban areas defined by their post-colonial and sectarian histories (Maputo, Beirut), and iconic 'world cities' (Barcelona, Berlin, Delhi and Rio de Janeiro). Such diversity is valuable because it emphasises the importance of understanding 'energy transition' in multiple dimensions: the chapters in this collection affirm time and again how struggles to craft more sustainable energy futures are about considerably more than delivering 'low carbon' or implementing climate governance. Through their grounded engagement with urban politics in cities in Europe, Africa, South Asia, Latin America and the United States, the book's chapters also show how urban (energy) histories matter, as infrastructural and political legacies produce different starting points for contemporary processes of energy transition. In much of Eastern Europe, for example, grid politics are marked by a history of universal (and largely free) access during the communist period and subsequent efforts to liberalise energy markets. In much of the global South, however, universal electricity access is only a recent aspiration and the extension of infrastructure is occurring through the combined actions of private and public actors, and formal and informal mechanisms. Urban energy politics, then, are profoundly plural.

For over a century electricity has been a potent technology for imagining urban futures, but what kinds of political worlds do urban electrical grids create? More specifically, under what conditions can electrical networks contribute to more progressive urban political formations and, conversely, when do they entrench corporate and/or state power? The chapters in this important collection stimulate creative thinking about the diversity of contemporary struggles over urban energy service provision around the world, and the geographies of energy transition to which they are giving rise.

Gavin Bridge
Durham, August 2015

References

Bennett, J., 2005. 'The Agency of Assemblages and the North American Blackout'. *Public Culture* 17(3): 445–65.

Cumbers, A., 2012. *Reclaiming Public Ownership: Making Space for Economic Democracy.* London and New York: Zed Books.

Graham, S. and S. Marvin, 2001. *Splintering Urbanism: Networked Infrastructures, Technological Mobilities and the Urban Condition.* London: Routledge.

Hughes, T., 1983. *Networks of Power: Electrification of Western Society, 1880–1930.* Baltimore: Johns Hopkins University Press.

IEA, 2010. *World Energy Outlook.* Paris: International Energy Agency.

IEA, 2014. *World Energy Investment Outlook, Special Report.* Paris: International Energy Agency.

McDonald, D. (ed.). 2009. *Electric Capitalism: Recolonising Africa on the Power Grid.* London and Cape Town: HSRC Press.

Mitchell, T., 2011. *Carbon Democracy: Political Power in the Age of Oil.* London and New York: Verso.

Platt, H., 1991. *The Electric City: Energy and the Growth of the Chicago Area, 1880–1930.* Chicago: Chicago University Press.

1   Introduction

Andrés Luque-Ayala and Jonathan Silver

This is a book concerned with the importance of electricity in the socio-political making of urban worlds. In what is increasingly described as an 'urbanized world' (UN-Habitat, 2010), an 'urban age' (Burdett & Sudjic, 2007) and the 'planeterization of the urban' (Brenner, 2013), everyday politics are shaped and defined through access to infrastructure systems and essential services. The electricity grid is, we argue, one of the key constituents of the urban, shaping the city's spatial, social and political landscapes as much as the ability of urban governance actors to respond to emerging global challenges. Across the world, cities are struggling with issues of energy security, climate change and resource constraints. Whilst cities in the global South endeavour to provide safe electricity access to growing urban populations, achieve cost recovery and address widespread clandestine connections, counterparts in the global North contend with addressing energy (and carbon) intensity and securing supply. In both cases, the urban energy landscape is increasingly characterised by rising consumption, the loss of affordability, growing energy poverty and the inability to sustain flows of electricity into households. As local governments and populations respond to these challenges, championing or demanding a variety of socio-technical transformations in local energy infrastructures, we contend that a new urban politics is unfolding.

The book examines how electricity networks are increasingly coming to the forefront of contemporary debates on urban politics. Its objective is to foreground questions concerning (in)equality, domination and control, citizenship and the (re)configuration of power in towns and cities, and how this is achieved (or undermined) through an engagement with energy infrastructures – specifically, electricity. This is what we see as the (political) *Geographies of the Electric City*, central – we would contend – to an understanding of contemporary processes of urbanisation. The premise of this book is a rejection of the view that electricity remains merely in the realm of the technological/technical or in the domain of planners and engineers, a perspective that has marginalised our understanding of the role of material infrastructures in shaping the social and political landscapes of the urban. In examining such a form of 'electropolis', by joining the very materiality of electricity – its flows, cables, meters and pylons – with the Greek notion historically used for understanding the city as a political community, we seek to foreground the politics that emerge from interacting with and through urban energy networks. Thus, the geographies of the electric city do not simply denote a dependence of both city and citizens on electricity; rather they signal the co-constitution of city and citizenship through the grid. In line with this, the book sets out to examine the political nature of electricity networks through the lens of nine urban regions around the world: the American South (Rocky Mount, in the United States), Plovdiv (Bulgaria), Rio de Janeiro (Brazil), Delhi (India), Maputo (Mozambique), Berlin (Germany), Beirut (Lebanon), Barcelona (Spain) and Athens (Greece).

In the world of electricity, the city has the unique condition of being a highly dense node of consumption. In the Western world, the early history of electricity is by and large an urban history (Hughes, 1983; Tarr, 1984; Nye, 1992). Over the course of the 20th century, as electricity grids developed in and around towns and cities worldwide, electricity became an essential component of modern urban life. Even today, access to modernity is measured by the extent to which populations have access to electricity, a fundamental way in which the politics of energy is constituted. As contemporary modernity arguably takes an urban shape, and demand for energy grows, electricity has become central to a global urbanisation process. Over the past decade the world's urban population surpassed 3.5 billion, and is predicted to rise to 6.3 billion by the middle of the 21st century (UN-DESA, 2015). These energy-dependent urban dwellers – and their rapidly growing towns and cities – drive global resource and finance flows and structure energy-intensive modes of production and consumption. The International Energy Agency (2009) suggests that by 2030 urban populations will account for around 75 per cent of global energy demand, with non-OCED countries accounting for 80 per cent of the projected increased demand above 2006 levels. Such vast, planetary scale urbanisation rests on carbon-based energy sources (Mitchell, 2011), often delivered through electricity networks which stretch and weave across city space to sustain the foundations of everyday urban life. Yet on a global scale, as of 2011, 1.3 billion people still lack access to electricity (IEA, 2013). There are of course significant differences in access depending on income levels, as only 23 per cent of the population in low-income countries have access, compared with 81.5 per cent in middle-income countries (World Bank, 2013). There are still further differences in relation to rural and urban populations. As of 2009, only 68 per cent of the world's rural population had electricity access compared to 93.7 per cent in urban areas (IEA, 2011).

So if electricity access in urban regions remains significantly higher than in rural areas, why is electricity in cities still a political matter? As a whole, energy systems, as 'forms of socio-technical organisation, financial circulation and political power', are deeply involved in the opening or closing of democratic possibilities of both historic and contemporary societies alike (Mitchell, 2009: 399). In the case of electricity, it is the very possibility and condition of access and control of such types of essential services that determines a condition of (in)equality and the full recognition of rights and citizenship. In the global South, as explored through the contributions of this book, the energy and development debate goes significantly beyond issues of access. It spans across a multiplicity of other domains such as affordability, reliability, rationalities around universal service provision, equality and difference, and the constitution of particular kinds of cities, state-society relations and forms of citizenship. For many of those living in cities in the global South, electricity access often occurs in the form of irregular, patchy and informal connections. Often deemed illegal, these connections are undersized for the amount of energy they are required to deliver, with constant voltage variations resulting in frequent supply interruptions and damaged domestic appliances. They are characterised by improvised equipment, an absence of safety standards, deadly fires and a multitude of health hazards. As Edgar Pieterse articulates in his examination of 'the right to the city' (2008), resources – such as energy, water, shelter and waste management – enable the right to life, and lie within the frame of civil and political rights, or what is also known as first generation rights. In calling for complexity and power as privileged lenses to understand the city, Pieterse (2008) suggests the need for an equal engagement with the managerial, the technical and the political dimensions associated with the city's technical and infrastructural dynamics. In doing so he points to urban inequality as the central problematic of our age that needs to be addressed.

Electricity in cities in the global North is equally political. With access levels nearing 100 per cent, these cities experience a relatively even access to service provision across populations. Whilst increasingly splintered (Graham & Marvin, 2001), it is a form of service provision which has long acknowledged the role that essential services play in the constitution of equal citizenship. The contributions of this book suggest how the electricity grid in cities in the global North remains a site of constant conflict and contestation: towards the maintenance of affordability in the context of urban austerity; towards the constitution of equal race relationships; towards the adoption of progressive ideas around environmental sustainability; and in the mobilisation of a new role for the public in the ownership, control and management of utility services.

Through an analysis of geographically diverse case studies, this edited volume examines how the configuration of electricity networks opens up multiple political debates that foreground the contested nature of electricity systems across towns and cities. The collection makes an explicit attempt to bridge North–South divides, looking at a diversity of large and medium-size urban areas across the world. By examining the uneven power relations embedded in the make-up of electricity networks in urban areas, the chapters bring together the insights of urban infrastructure studies (traditionally focused on water and sanitation networks) and the emerging field of energy geographies. They establish how the electricity grid operates as a critical political site for mediating urban life whilst opening up new possibilities for expressing dissent and advancing social justice, democratic service delivery and urban autonomy. The point of this collection is not simply to uncover the political, but to illustrate how insight and inspiration towards more progressive forms of service delivery can be drawn from these examples. This is done by outlining a social research agenda around electricity that recognises the – often historical – uneven geographies of urban electricity networks, the political nature of contemporary attempts to 'rewire' and transform them, and the increasingly common use of protest and resistance by residents and social movements alike to express discontent around service provision and seek new forms of access. We hope that our collective effort as editors and contributors of this volume, through seeking to capture this global landscape of *Energy, Power and Protest on the Urban Grid,* will enrich the steadily growing sub-field of study that takes seriously the role of energy in mediating urban life.

In this introductory chapter we examine the broad political nature of urban infrastructures, as a prerequisite for a more in-depth examination of the imperatives surrounding urbanisation and electricity, together with the politics that emerge from such processes. We also suggest that an urban politics of the grid is constituted through three broad domains. First, through the *uneven geographies of urban energy networks,* which point to the historical and contemporary production of racial and social inequalities through and by the electricity grid. Second, through attempts towards *rewiring the urban grid,* where dominant urban energy configurations are sought to be transformed in response to perceived global challenges – such as climate change, resource constraints and limited energy security – and in search of a more progressive configuration of resource access, ownership, control and management. Third, through the emergence of *social movements and protest in and around urban electricity,* where ongoing and multiple forms of resistance operate through the grid itself as a mechanism to challenge and transform the dominant structures of power in the city.

Infra-Politics!

A first step in framing and conceptualising the geographies of the electric city lies in an understanding of the political nature of urban infrastructures. Networked infrastructures shape spatio-political dynamics in the city, configuring understandings of citizenship as well as relations of control, inequality and differentiated access. Between 1880 and 1960 electricity networks in the Western world were instrumental in the emergence of the modern ideal of service access and urban cohesion (Hughes, 1983; Nye, 1999). The integrated networked city, where electricity, gas, transport and communication networks readily service all citizens, became the standardised norm and the dominant discourse in urban planning. This mobilises reason and democracy towards particular urban forms and the fulfilment of collective goals (Graham & Marvin, 2001). Such universal and democratic understandings of service access were to change in the second half of the 20th century, when the privatisation and liberalisation of utilities resulted in processes of 'urban splintering' rather than integration. Electricity and other infrastructure networks, seen earlier by planners, engineers and architects as the integrators of urban space, were reconfigured as specialised, privatised and customised, providing increased connectivity to some while bypassing others. The resulting dynamics encouraged the fragmentation of the socio-material fabric of the city whilst exacerbating spatial segregation and social polarisation, raising questions about the existence of the city as a unitary whole (Graham & Marvin, 2001). Infrastructure connections and access were shown to have an uneven distribution in response to different political and economic interests and capacities, alongside what Graham and Marvin (2001: 9) describe as a 'paradoxical trend towards the reinforcement of local boundaries'.

In the global South, where the realisation of a modern infrastructural ideal has been highly uneven (Coutard, 2008), urban infrastructures have also played a role in generating and maintaining fragmentation, injustice and exclusion. Southern cities have always been characterised by fragmented urban fabrics and infrastructures in a permanent state of disrepair and improvisation (Graham & Thrift, 2007; McFarlane & Rutherford, 2008). Here the development of infrastructure networks has been 'socially constructed by various interest groups through an array of tensions, tactics and complexities, which are far more problematic for (just and equitable) infrastructure provision than any technical issues' (McFarlane & Rutherford, 2008: 370). Characterised by 'incomplete modernities', and with service provision historically concentrated on the wealthy and elites (Gandy, 2006; Zérah, 2008), infrastructure in cities in the global South can be seen as 'splintered' rather than 'splintering' (Kooy & Bakker, 2008: 1843). Rather than based on a single unified network, infrastructure here is characterised by a multiplicity of – formal and informal – provision mechanisms that make up 'spatially separated but linked "islands" of networked supply' (Bakker, 2003: 337).

Whilst research asserting the political nature of urban infrastructures has long been established across urban infrastructure studies (Graham & Marvin, 2001; McFarlane & Rutherford, 2008) and urban political ecology (Bakker, 2003; Swyngedouw, 2004; Kaika, 2005), this work has been realised primarily through analyses of water and sanitation networks (for notable exceptions engaging with electricity, see Graham, 2000; Swyngedouw, 2007; Jaglin, 2008; Moss, 2009a, 2009b). In recent years, a growing body of research has started to focus on the specificity of electricity in the configuration of urban politics (Monstadt, 2009; Bulkeley et al., 2014). Several of these studies focus on issues of social and environmental justice in the context of the commercialisation of electricity and the neoliberal restructuring of the sector (McDonald, 2009). Others have examined the politics of urban electricity emerging through climate change responses, critically evaluating a process of reconfiguring urban infrastructures towards a low-carbon transition (Rutherford & Coutard, 2014; Bulkeley et al., 2014). In order to further conceptualise this emerging field, we propose three ways in which an urban politics on the grid can be understood.

Conceptualising Urban Politics on the Grid

Since the turn of the 21st century, scholars from within science and technology studies, sociology, geography and anthropology have advocated for the emergence of an energy social science (Guy & Shove, 2000; Rohracher, 2008; Bridge et al., 2013; Strauss et al., 2013). Intellectually disconnected from the perspectives examined in the previous section, this literature asks energy researchers to venture beyond the 'ready made but limited roles [of] addressing the so-called "human dimensions" of energy efficiency' (Guy & Shove, 2000: 2). It asks questions around politics and justice in relation to energy systems and their transformation (Buzar, 2007; Walker, 2008; Bickerstaff et al., 2013) whilst examining how energy processes and the transformation of networked systems reconfigure socio-spatial dynamics (Bridge et al., 2013).

Arguably, within this emerging field, the urban scale has rarely figured as a primary site of analysis (Bridge et al., 2013). Yet, bringing the urban into a social analysis of energy flows points to an energy problem that is no longer a matter of national security. This stands in sharp contrast with a politics of energy that is primarily defined within a geopolitical framework, framed by issues of national security and concerned with global flows (Yergin, 2006; Mitchell, 2008; Bradshaw, 2009). Rather, here the energy problem emerges out of concern with different forms of security at local levels, from securing urban growth and local resource availability (Luque-Ayala, 2014) to reducing social vulnerability and increasing energy affordability at the household level (Buzar, 2007; Walker, 2008; Walker & Day, 2012). Here the role of energy in building communities is foregrounded, alongside the energy practices, assemblages and landscapes shared by members of these communities in their everyday life (Castán Broto et al., 2014). We argue that an analysis of the urban – of its everyday materiality and the scalar dynamics at the interface between infrastructures and novel forms of urban governance (Bulkeley, 2005; Bulkeley & Castán Broto, 2013; Bulkeley & Betsill, 2013) – provides significant contributions towards further examining both the politics of energy and the politics of infrastructures.

Earlier, we pointed out how the urban politics on the grid travels significantly beyond issues of access. Looking at urban energy dynamics through the lenses provided by the very materiality of infrastructures illustrates hidden forms of political rationality alongside different forms of constituting the political (Larkin, 2013). As such, it is a reminder of the need to understand infrastructure as 'integral to the conduct of politics' (Barry, 2013: 2). A form of 'electric politics' is exercised, for example, through the power of connection and disconnection, through the use of the grid for the creation and/or consolidation of markets, and through the attempts of different stakeholders (from local communities to foreign powers) to obtain control and decision-making power over local resources. Once connection is in place, the threat of disconnection may be mobilised in order to govern conducts (Luque-Ayala, 2016) and colonise territories (Jabary Salamanca, 2011). The grid becomes a material technique for incorporating populations into economic markets and circuits of capital whilst promoting rational, economic and efficient behaviours around resource use (Luque-Ayala, 2016; Babourkova, this volume; Pilo', this volume). That very same grid is increasingly a contested space, disrupted by different stakeholders to advance their rights and social values – from labour rights (Verdeil, this volume) to environmental values (Maassen, this volume) – and targeted as the appropriate space from where a progressive politics may unfold – for instance via a re-municipalisation of electricity infrastructures, or through novel collective forms of infrastructure ownership (Colell and Neumann-Cosel, this volume). The remainder of this introduction outlines three conceptual ways of examining the constitution of the urban politics on the grid, or what we termed the geographies of the electric city – *uneven geographies, rewiring,* and *social movements and protest.* These three broad domains also serve to structure the three parts of the book.

The Uneven Geographies of Urban Energy Networks (Part I)

Electricity mediates power and inequality within the city. In the cities of the global South, for example, urban infrastructure studies point to how splintered infrastructural configurations and the exclusion from essential resource flows have been enacted from early colonial times as tools for segregation and domination (Kooy & Bakker, 2008; McFarlane, 2008; Silver, 2016). In a similar way in the global North, both historical and contemporary analyses of electricity access in cities uncover issues of racial inequality and segregation, or the advance of unequal racial relations under broad projects of social and demographic control (Harrison, this volume; Babourkova, this volume). Varying levels of service provision across households (such as informal, basic or premium), along with differential qualities and quantities (from patchy and irregular to reliable and safe), signal the production of uneven energy across neighbourhoods, towns and cities. Such differentiated service is as much an expression of structural inequalities – from rich and poor to dominant and subjugated – as a mechanism for the very formation and reproduction of these uneven geographies.

An examination of the ways in which the uneven geographies of electricity are produced can be drawn from both historical and contemporary perspectives. As Harrison illustrates through a case study of the American South in the late 19th century (this volume), variable electricity access plays a role in the production of race-based socio-spatial differentiation, in this case endorsing a vision of white supremacy and a collective position of racial antagonism. In a case study of Lagos, Gandy argues that the use of a historical perspective for uncovering uneven access to infrastructures provides analytical links between the past and the present, 'revea[ling] how structural factors operating through both the colonial and post-colonial periods have militated against any effective resolution to the city's [contemporary] worsening infrastructure crisis' (2006: 371). Uncovering an uneven geography of electricity networks needs to pay attention to the way in which political economies have shaped broader uneven and splintered access across the urban during the last century (Graham & Marvin, 2001). Through this, the overlapping modes of governance (colonialist, nationalist, neoliberal and so forth) that intersect and reshape these uneven geographies are revealed, teasing apart the localised histories of urban electricity grids (Silver, 2016).

The construction of uneven geographies of electricity is often traversed by attempts to establish, control and benefit from financial and monetary flows around resource provision. In the American South (Harrison, this volume), white industrial elites sought to profit from a particular racial and financial configuration of the electricity grid, as public investment in the development of a local grid allowed them to avoid tying their own capital to technologies required for industrial modernisation and business expansion. In Palestine (Jabary Salamanca, 2011), dispossession (of electric resources) as well as disconnection and reconnection (from the grid) have operated as tactics towards the consolidation of capitalist-colonialist modes of ownership and the capture of capital flows. Here electrification becomes central to both settler colonialism and the consolidation of a capitalist mode of production. In Plovdiv (Babourkova, this volume), premium electricity services for Roma minorities are mobilised in an attempt to ensure that users subscribe to a fee-paying logic – a neoliberal (and racialised) rationality embodied in the expansion of energy markets and constructed materially through the grid. Yes, profit making and the control of financial flows are seldom the primary drivers of an uneven geography of electricity grids. Rather, these processes, alongside the mobilisation of capital and the development of a variety of financial structures around the electricity grid, are about the consolidation of specific political rationalities.

Rewiring the Urban Grid (Part II)

The second broad domain by which an urban politics on the grid is constituted is through the purposeful attempts – by local governments as well as other state and non-state actors – to transform urban energy systems in response to perceived global challenges, such as climate change, resource scarcity and energy security (Hodson & Marvin, 2009). Here, a reconfiguration of energy systems operates as a way to ensure what Bridge et al. (2013: 331) describe as 'the availability and accessibility of energy services in a carbon-constrained world'. Responses to these challenges are being undertaken through socio-technical interventions which, by re-wiring energy systems, in effect 'rewire' the city, through new technologies and innovation (Geels, 2010), policy development (Jaglin, 2008) and an experimentation in governance modes (Hodson & Marvin, 2010; Bulkeley & Castán Broto, 2013).

However, re-wiring is never neutral or purely material. It is entangled with the advancement of specific rationalities – neoliberal, calculative, formalising. Constraints in resources and the push towards greater forms of sustainability are not the only drivers at play; the re-wiring is a socio-material expression of specific political rationalities that are reshaping power relations in the city. Re-wiring the urban electricity grid is, in effect, transforming historic socio-spatial configurations, through the adoption of different infrastructural and socio-technical arrangements. In doing this, it is shaping a new material politics for both city and network. In the global South, for example, this re-wiring of the urban grid is also conceived as a response to the uneven geographies examined earlier, particularly the need to reduce non-technical losses whilst increasing and improving energy access for the poor, securing both affordability and payment capacity and ensuring the grid's financial stability through the creation of new energy markets (Pilo', this volume; Baptista, this volume). In Rio de Janeiro, low-income energy consumers are reconfigured as customers after decades of accessing electricity via informal networks. Forced to pay for electricity for the first time, *favela* dwellers experience both an improvement in service and a reduction in domestic budgets (Pilo', this volume). Similarly, in Maputo, prepayment electricity meters empower users to access reliable energy services at their will. Such prepayment strategies, in disregarding the needs of the poor, naturalise the failure of the modern infrastructure ideal – to be connected to the grid is no longer a guarantee of energy access (Baptista, this volume). As illustrated by the case of Delhi, service access and infrastructural planning are no longer governed by the public sector but by the commercial imperatives of a liberalised electricity sector (Criqui, this volume). As the call for greater access unfolds, political rationalities expand and consolidate.

'Re-wiring' urban electricity is a complex, often contested, endeavour characterised by multiple social interests and different overlapping agendas (Hodson & Marvin, 2007). Efforts leading to such interventions often build on existing systems through attempts to improve infrastructural resilience (Braun, 2014), a reduction in the carbon content of both energy infrastructures and urban development modes (While et al., 2010; Castán Broto & Bulkeley, 2013), and the establishment of novel ways of incorporating local concerns and circulatory flows within multi-scalar energy systems (Luque-Ayala, 2014). But this re-wiring also challenges traditional grid arrangements through the promotion and adoption of different political and user configurations. When unpacking the unfolding energy politics across contemporary processes of 're-wiring', it is important to consider the multi-scalar nature of cities and the relational nature of electricity networks (Bulkeley, 2005: Silver, 2016). From consumption practices by single dwellers to neighbourhood-level distribution networks and regional transmission grids, electricity is never exclusively urban. The urban governance of electricity, like that of other resource flows, is shaped and influenced by a series of global, international and regional circulations of policy, finance and technologies across local contexts (Bulkeley et al., 2014). Maintaining this wider focus is important as following Swyngedouw and Heynen (2003: 912), 'these dynamics are embedded within networked or territorial scalar configurations that extend from the local milieu to global relations'. Responding to this scalar condition demands an analysis of a multiplicity of sites that may be navigated by municipalities, utility companies and users as well as the current institutional arrangements that incorporate stakeholders from within and outside the city – a shifting collection of actors.

Social Movements and Protest in the Electric City (Part III)

Finally, an urban politics on the grid is configured through novel political rationalities and logics of control as much as through attempts to resist them. The material included in this book shows that understanding the contemporary geographies of the electric city requires positioning resistance at the heart of the governing and politics equation. Drawing on a Foucauldian approach where 'the final word on power is that resistance comes first' (Deleuze, 1988: 74), unpacking the contemporary politics of infrastructures demands providing greater attention to what Legg (2011: 128) calls 'subaltern experience, refusal, autonomy and resistance'. Resistance provides a powerful conceptual entry point for what Foucault (1979) terms 'a new economy of power relations': analysing power not through the operations of institutions, but through the forms of resistance and antagonisms against such power. Here, resistance is not exterior to power but located at all points within the network; power can only operate in relation to such opponents, points of support and targets (Foucault, 1979).

Politics, protest and resistance within energy networks come across strongly through an intense awareness of the technical and material nature of energy politics. Timothy Mitchell's (2009) seminal study comparing the politics and democratic potential of oil and coal sets out ways of thinking about energy politics through material and technical configurations. In his work, protest and resistance co-configure the energy system. This is echoed by Barry (2013), for whom disputes around energy resources, particularly oil and its global pipelines, take both material and performative forms. Within urban energy configurations, intervening in and through the materiality of infrastructures is an important strategy for often marginalised urban dwellers in resisting conditions of inequality. Silver (2014), drawing on postcolonial debates in urban studies, examines how urban dwellers in Accra bypass formal energy connections in order to create free circuits of energy and prefigure future and potentially emancipatory infrastructural conditions. Material configurations play a similar role in Cupples' description of resistance to electricity privatisation in urban Nicaragua, as she draws on Actor-Network Theory to position the electricity meter, the Internet and other media spaces as key agents of resistance (Cupples, 2011).

The local politics of energy evolves from ongoing and multiple forms of resistance to the dominant forms of power that are constituted through the grid itself. Protest and other forms of social mobilisation are transforming how we conceive and access electricity, whilst new energy configurations are advancing a re-localisation of the power of the grid in the hands of a variety of urban social interests. Through forms of contestation, protest and resistance carried out by individuals as well as an assortment of civic organisations, social movements and public mobilisations, these emerging grid configurations challenge both the uneven production of the network and the attempts to rewire these systems. As in the case of water infrastructure, where activists have explored strategies and tactics to challenge dominant understandings of water as a commodity and promote alternative governance models in response to water privatisation (Shiva, 2002; Bakker, 2007), social movements are seeking to regain control of electricity networks. Echoing Bakker's understanding of the tensions, potential and experiences of the 'commons versus the commodity' in the struggle to access and control water infrastructures (2007), we can draw important lessons from how grassroots pressures and actions may challenge inequality and segregation across the urban grid. In Athens, for example, protest reveals electricity access as a fundamental common good, sparking widespread resistance against new forms of taxation (Alexandri and Chatzi, this volume). In Berlin, the reconstruction of local power through forms of cooperative ownership of the local electricity utility distribution system is aimed at involving local population within decision-making structures, and in this way reclaiming energy infrastructures for communities (Colell and Neumann-Cosel, this volume). This grid-embedded resistance also occurs when actors – such as utility workers or environmental activists – 'try to reap advantage from the technical-political vulnerability of urban electricity circuits' in order to advance their own agendas (Verdeil, this volume; see also Maassen, this volume). Here the electricity grid is both a vehicle for resistance and a privileged space where urban politics occur. Examining grid resistance in action raises questions around the competing rationalities through which the grid should and could be governed, from public, private and communal perspectives to market rationalities, solidarity and co-operativism. The protest and social movements around urban electricity, in effect, regain wider urban power through the network itself.

Detailed Outline

Each of the nine chapters of this book is based on an empirical case study that provides conceptual tools towards substantiating and expanding the three suggested ways of examining the constitution of urban politics on the grid proposed above. *Chapter 2*, by Conor Harrison, examines the role of race in the early configuration of urban electricity grids in the American South through a case study of Rocky Mount, North Carolina, in the United States. The chapter considers the historical-geographical emergence of electricity networks, focusing on how uneven processes of urbanisation and electrification interact with two contradictory planning ideologies: the Progressive Era reformist movement and the Jim Crow laws that governed racial segregation in Southern cities. With a segregation-led division of different populations on the one hand, and a simultaneous quest for efficiency and order on the other, these ideologies combine to shape infrastructural politics in the American South in ways that continue to shape a racialised geography of electricity.

Further unpacking the role of race in the production of uneven geographies of urban electricity networks, Rosalina Babourkova in *Chapter 3* examines the broader social tensions around service provision that emerge in the overlap between positive and negative forms of discrimination towards ethnic minorities. Focusing on Roma settlements in Plovdiv, Bulgaria, the chapter explores the differentiated utility provision in this post-socialist city. On one hand, a long history of political promises stating that the Roma do not have to pay for electricity – given their disadvantaged condition – leads to a naturalisation of non-payment, reinforcing a widespread perception of the Roma as illiterate, lacking discipline and respect for legal and social norms. On the other hand, mediated through the arrival of the private sector in utility provision, pervasive non-payment of electricity in Romani neighbourhoods have led to increased disconnections, electricity rationing and discriminatory metering practices. Through a novel methodology based on an analysis of comments provided by readers of online press coverage, Babourkova examines the resulting subtle yet audible discursive politics around race, ethnicity and access to electricity.

Francesca Pilo', in *Chapter 4*, looks at the regularisation of electricity provision in the *favelas* of Rio de Janeiro, Brazil. The chapter examines the transformation of informal and unmetered consumers – accessing unpaid electricity by means of illegal connections – into metered customers and 'responsible paying consumers'. Here Pilo' unpacks the limits and paradoxes of technologies and policies producing new electricity markets and forms of subjectivity for the urban poor. Through an analysis of the technical interventions and social techniques used by private utility providers (for instance, anti-theft equipment, progressive billing and energy education), the chapter uncovers an emerging normativity in the consumption practices of low-income users operating in accordance with market logics. This process, highly traversed by energy efficiency rationalities, targets 'those who are actually far from being the biggest users of electricity' (Pilo', this volume: 83).

Chapter 5 focuses on the emerging urban role of private electricity distribution companies in Delhi, India, and the way these organisations bypass the structural limitations of the public sector to establish new modes of resource access and efficient forms of service delivery. Here, Laure Criqui examines the messy overlap between formal and informal planning practices in Delhi, and the technical and legal challenges of operating in informal settlements. Through bypassing the tension between legal and illegal, formal and informal, private utility companies (re)incorporate informal settlements into the city through infrastructural expansion.

In *Chapter 6*, Idalina Baptista examines prepaid electricity systems in Maputo, Mozambique, to argue for an expanded theorisation of prepayment as a shaper of societal politics beyond economic relations. In using an ethnographic sensibility towards the practices of everyday consumption of electricity in peri-urban areas, the chapter investigates how the prepayment meter, as a material and symbolic object, conjures diverse opportunities for individuals to articulate their understanding of what being a 'citizen' means and what is expected (or not expected) of the state.

Chapter 7, by Arwen Colell and Luise Neumann-Cosel, looks at the emergence of social movements demanding a more proactive stance in grid operations and a voice in the management of local infrastructures. Going beyond the calls for the re-municipalisation of public services, civil society movements in Germany are seeking a direct role in grid management through citizen-led cooperatives. Through an analysis of the proposed 're-communalisation' of Berlin's distribution utility company, the chapter explores the potential of cooperative organisational forms for transforming the politics of urban electricity. In Berlin, BürgerEnergie aims to develop the required social and financial capital for putting the city's grid under citizens' control. This chapter, written by academics and activists involved in the process, discusses the opportunities and potential of this transformative mode of service provision.

Chapter 8, by Eric Verdeil, discusses the role of protest and contestation within and through the power grid. Drawing on the experience of Beirut, Lebanon, the chapter uncovers the various powers – major and minor, formal and informal – that disrupt and reconfigure urban politics through their ability to connect and disconnect electricity flows. Beirut's electricity infrastructure, significantly affected by Lebanon's Civil War (1975–1990) and by the 2006 Israel–Hezbollah War, supplies power, on average, for only half of the day. Verdeil refers to Beirut as 'a metropolis of darkness', where two million inhabitants cope on a daily basis with extended periods of power failure. In recent years the poor condition of the city's power infrastructure has led to an increasing number of street protests, as citizens are forced to buy electricity from small-scale informal and privately run 'micro-grids' running on diesel generators. Through developing a typology of these novel grid politics and their ability to disrupt and alter the power relations of the city, the chapter suggests some important theoretical considerations for understanding protest within and across energy systems.

Anne Maassen, in *Chapter 9*, examines two contrasting initiatives advancing the uptake of solar energy within Barcelona's electricity grid. The chapter provides an exploration of competing visions on alternative urban energy relationships – of 'eco-empowerment', enacted by a set of Barcelona-based activists known as the solar guerrillas, and 'carbon reduction', performed by the municipal energy agency (and by extension, the Municipality of Barcelona). While both effectively alter an undesirable status quo, differences in these discourses and their associated socio-material practices draw attention to wider contestations across unfolding civic and electric futures.

The final empirical chapter, *Chapter 10*, examines the interface between energy taxation and a politics of austerity in Athens, Greece. In the context of a wide-spread attack on the welfare state in the city, Georgia Alexandri and Venetia Chatzi investigate forms of social resistance and mobilisation against the loss of energy affordability. Their chapter examines the rising neighbourhood mobilisations call-ing for the non-payment of new energy taxes and the social calls against power cuts to low-income households. The authors point to what could be termed as an emerging energy austerity. In becoming the focus of social mobilisation and pro-test against austerity, energy operates as a catalyst of civil disobedience and new social formations.

We suggest that the three broad domains of urban politics on the grid proposed in this collection – the examination of the *uneven geographies of urban energy networks,* the analysis of attempts to *rewire the urban grid* and the consideration of *social movements and protest* in the politics of urban infrastructures – outline a future research agenda for both energy social science and urban studies; an agenda around the (political) *Geographies of the Electric City,* or the co-constitution of cities and the political positions and struggles of their citizens through energy infrastructures. This is an agenda that seeks to open up a diverse and pluralised series of conceptual and methodological approaches to understand energy politics beyond the narrow confines of technological concerns, contributing to debates about the primarily socio-political nature of urban infrastructures.

References

Bakker, K., 2003. 'Archipelagos and Networks: Urbanization and Water Privatization in the South'. *Geographical Journal* 169: 328–41.
Bakker, K., 2007. 'The "Commons" Versus the "Commodity": Alter-globalization, Anti-privatization and the Human Right to Water in the Global South'. *Antipode* 39(3): 430–55.

Barry, A., 2013. *Material Politics: Disputes Along the Pipeline.* Chichester: Wiley Blackwell.

Bickerstaff, K., G.P. Walker and H. Bulkeley (eds.), 2013. *Energy Justice in a Changing Climate: Social Equity and Low-Carbon Energy.* London: Zed Books.

Bradshaw, M.J., 2009. 'The Geopolitics of Global Energy Security'. *Geography Compass* 3(5): 1920–37.

Braun, B., 2014. 'A New Urban Dispositif? Governing Life in an Age of Climate Change'. *Environment and Planning D: Society and Space* 32: 49–64.

Brenner, N., 2013. 'Theses on Urbanization'. *Public Culture* 25(169): 85–114.

Bridge, G., S. Bouzarovski, M. Bradshaw and N. Eyre, 2013. 'Geographies of Energy Transition: Space, Place and the Low-carbon Economy'. *Energy Policy* 53: 331–40.

Bulkeley, H., 2005. 'Reconfiguring Environmental Governance: Towards a Politics of Scales and Networks'. *Political Geography* 24: 875–902.

Bulkeley, H., and M.M. Betsill, 2013. 'Revisiting the Urban Politics of Climate Change'. *Environmental Politics* 22 (1): 136–54.

Bulkeley, H. and V. Castán Broto, 2013. 'Government by Experiment? Global Cities and the Governing of Climate Change'. *Transactions of the Institute of British Geographers* 38(3): 361–75.

Bulkeley, H., A. Luque-Ayala and J. Silver, 2014. 'Housing and the (Re)configuration of Energy Provision in Cape Town and São Paulo: Making Space for a Progressive Urban Climate Politics?' *Political Geography* 40: 25–34.

Burdett, R. and D. Sudjic, 2007. *The Endless City: An Authoritative and Visually Rich Survey of the Contemporary City.* London: Phaidon Press.

Buzar, S., 2007. 'The "Hidden" Geographies of Energy Poverty in Post-socialism: Between Institutions and Households'. *Geoforum* 38(2): 224–40.

Castán Broto, V. and H. Bulkeley, 2013. 'A Survey of Urban Climate Change Experiments in 100 Cities'. *Global Environmental Change,* 92–102.

Castán Broto, V., D. Salazar and K. Adams, 2014. 'Communities and Urban Energy Land-scapes in Maputo, Mozambique'. *People, Place and Policy* 8: 192–207.

Coutard, O., 2008. 'Placing Splintering Urbanism: Introduction'. *Geoforum* 39(6): 1815–20.

Cupples, J., 2011. 'Shifting Networks of Power in Nicaragua: Relational Materialisms in the Consumption of Privatized Electricity'. *Annals of the Association of American Geographers* 101: 939–48.

Deleuze, G., 1988. *Foucault.* Minnesota: University of Minnesota Press.

Foucault, M., 1979. *The History of Sexuality Vol. I: An Introduction.* London: Allen Lane.

Gandy, M., 2006. 'Planning, Anti-planning and the Infrastructure Crisis Facing Metropoli-tan Lagos'. *Urban Studies* 43(2): 371–96.

Geels, F., 2010. The Role of Cities in Technological Transitions. In: H. Bulkeley, V. Castán Broto, M. Hodson and S. Marvin (eds.), *Cities and Low Carbon Transitions.* London: Routledge, 13–28.

Graham, S., 2000. 'Constructing Premium Network Spaces: Reflections on Infrastructure Networks and Contemporary Urban'. *International Journal of Urban and Regional Research* 24(1): 183–200.

Graham, S. and S. Marvin, 2001. *Splintering Urbanism: Networked Infrastructures, Tech-nological Mobilities and the Urban Condition.* London: Routledge.

Graham, S. and N. Thrift, 2007. 'Out of Order Understanding Repair and Maintenance'. *Theory, Culture & Society* 24(3): 1–25.

Guy, S. and E. Shove, 2000. *The Sociology of Energy, Buildings and the Environment: Constructing Knowledge, Designing Practice Vol. 5*. London: Psychology Press.

Hodson, M. and S. Marvin, 2007. 'Understanding the Role of the National Exemplar in Constructing "Strategic Glurbanization" '. *International Journal of Urban and Regional Research* 31(2): 303–25.

Hodson, M. and S. Marvin, 2009. ' "Urban Ecological Security": A New Urban Paradigm?' *International Journal of Urban and Regional Research* 33(1): 193–216.

Hodson, M. and S. Marvin, 2010. 'Can Cities Shape Socio-Technical Transitions and How Would We Know If They Were?'. *Research Policy* 9(4): 477–85.

Hughes, T., 1983. *Networks of Power: Electrification in Western Society, 1880–1930*. Baltimore: Johns Hopkins University Press.

IEA, 2009. *Cities, Towns and Renewable Energy: Yes in My Frontyard*. Paris: International Energy Agency.

IEA, 2011. *International Energy Outlook*. Paris: International Energy Agency.

IEA, 2013. *International Energy Outlook*. Paris: International Energy Agency.

Jabary Salamanca, O., 2011. 'Unplug and Play: Manufacturing Collapse in Gaza'. *Human Geography* 4(1): 22–37.

Jaglin, S., 2008. 'Differentiating Networked Services in Cape Town: Echoes of Splintering Urbanism?'. *Geoforum* 39(6): 1897–906.

Kaika, M., 2005. *City of Flows: Modernity, Nature, and the City*. London: Psychology Press.

Kooy, M. and K. Bakker, 2008. 'Splintered Networks: The Colonial and Contemporary Waters of Jakarta'. *Geoforum* 39(6): 1843–58.

Larkin, B., 2013. 'The Politics and Poetics of Infrastructure'. *Annual Review of Anthropology* 42: 327–43.

Legg, S., 2011. 'Assemblage/Apparatus: Using Deleuze and Foucault'. *Area* 43(2): 128–33.

Luque-Ayala, A., 2014. Reconfiguring the City in the Global South: Rationalities, Techniques and Subjectivities in the Local Governance of Energy [PhD]. *Geography*. Durham: Durham University.

Luque-Ayala, A., 2016. From Consumers to Customers: Regularizing Electricity Networks in São Paulo's Favelas. In: M. Hodson and S. Marvin (eds.), *Retrofitting Cities: Priorities, Governance and Experimentation*. London: Routledge, pp. 171–91.

McDonald, D., 2009. Electric Capitalism: Conceptualising Electricity and Capital Accumulation in (South) Africa. In: D. McDonald (ed.), *Electric Capitalism: Recolonising Africa on the Power Grid*. London: Earthscan. pp. 8–49.

McFarlane, C., 2008. 'Governing the Contaminated City: Infrastructure and Sanitation in Colonial and Post-colonial Bombay'. *International Journal of Urban and Regional Research* 32(6): 415–35.

McFarlane, C. and J. Rutherford, 2008. 'Political Infrastructures: Governing and Experiencing the Fabric of the City'. *International Journal of Urban and Regional Research* 32(2): 363–74.

Mitchell, C., 2008. *The Political Economy of Sustainable Energy*. New York: Palgrave Macmillan.

Mitchell, T., 2009. 'Carbon Democracy'. *Economy and Society* 38(3): 399–432.

Mitchell, T., 2011. *Carbon Democracy: Political Power in the Age of Oil*. New York: Verso.

Monstadt, J., 2007. 'Urban Governance and the Transition of Energy Systems: Institutional Change and Shifting Energy and Climate Policies in Berlin'. *International Journal of Urban and Regional Research* 31(2), 326–43.

Monstadt, J., 2009. 'Conceptualizing the Political Ecology of Urban Infrastructures: Insights from Technology and Urban Studies'. *Environment and Planning A* 41(8): 1924–42.

Moss, T., 2009a. 'Divided City, Divided Infrastructures: Securing Energy and Water Services in Postwar Berlin'. *Journal of Urban History* 35(7): 923–42.

Moss, T., 2009b. 'Intermediaries and the Governance of Sociotechnical Networks in Transition'. *Environment and Planning A* 41(6): 1480–95.

Nye, D., 1992. *Electrifying America Social Meanings of a New Technology, 1880–1940*. Cambridge MA: MIT Press.

Nye, D.E., 1999. *Consuming Power: A Social History of American Energies*. Cambridge MA: MIT Press.

Pieterse, E., 2008. *City Futures: Confronting the Crisis of Urban Development*. London: Zed Books.

Rohracher, H., 2008. 'Energy Systems in Transition: Contributions from Social Sciences'. *International Journal of Environmental Technology and Management* 9(2): 144–61.

Rutherford, J. and O. Coutard, 2014. 'Urban Energy Transitions: Places, Processes and Politics of Socio-technical Change'. *Urban Studies* 51(7): 1353–77.

Shiva, V., 2002. *Water Wars: Privatization, Pollution and Profit*. New Delhi: India Research Press.

Silver, J., 2014. 'Incremental Infrastructures: Material Improvisation and Social Collaboration Across Post-colonial Accra'. *Urban Geography* 35(6): 788–804.

Silver, J., 2016. 'Disrupted Infrastructures: An Urban Political Ecology of Interrupted Electricity in Accra'. *International Journal of Urban and Regional Research*. [Online] 8 January. DOI: 10.1111/1468-2427.12317.

Strauss, S., S. Rupp and T. Love, 2013. Powerlines: Cultures of Energy in the Twenty-First Century. In: S. Strauss, S. Rupp and T. Love (eds.), *Cultures of Energy: Power, Practices, Technologies*. Walnut Creek, CA: Left Coast Press, 10–38.

Swyngedouw, E., 2004. *Social Power and the Urbanization of Water: Flows of Power*. Oxford: Oxford University Press.

Swyngedouw, E., 2007. Impossible 'Sustainability' and the Postpolitical Condition. In: R. Krueger and D. Gibbs (eds.), *The Sustainable Development Paradox: Urban Political Economy in the United States and Europe*. New York: The Guilford Press, pp. 13–40.

Swyngedouw, E. and N.C. Heynen, 2003. 'Urban Political Ecology, Justice and the Politics of Scale'. *Antipode* 35(5): 898–918.

Tarr, J., 1984. The Evolution of Urban Infrastructure in the Nineteenth and Twentieth Centuries. In: E. Hanson (ed.), *Perspectives on Urban Infrastructure*. Washington, DC: National Academy Press, 4–65.

UN-DESA, 2015. *World Urbanization Prospects: The 2014 Revision* (ST/ESA/SER.A/ 366). United Nations Department of Economic and Social Affairs, Population Division. Available from: esa.un.org/unpd/wup/highlights/wup2014-highlights.pdf [Accessed 2 February 2016].

UN-Habitat, 2010. *State of the World's Cities 2010/11: Bridging the Urban Divide*. London: Routledge.

Walker, G., 2008. 'Decentralised Systems and Fuel Poverty: Are There Any Links or Risks?'. *Energy Policy* 36(12): 4514–17.

Walker, G. and R. Day, 2012. 'Fuel Poverty as Injustice: Integrating Distribution, Recognition and Procedure in the Struggle for Affordable Warmth'. *Energy Policy* 49: 69–75.

While, A., A. Jonas and D. Gibbs, 2010. 'From Sustainable Development to Carbon Control: Eco-state Restructuring and the Politics of Urban and Regional Development'. *Transactions of the British Institute of Geographers* 35(1): 76–93.

World Bank, 2013. *Data – Access to Electricity (% of Population)* [Online]. Available from: http://data.worldbank.org/indicator/EG.USE.ELEC.KH.PC [Accessed 21 January 2014].

Yergin, D., 2006. 'Ensuring Energy Security'. *Foreign Affairs* 85(2): 69–82.

Zérah, M.-H., 2008. 'Splintering Urbanism in Mumbai: Contrasting Trends in a Multi-layered Society'. *Geoforum* 39(6): 1922–32.

Part I

The Uneven Geographies of Urban Energy Networks

2   The American South
Electricity and Race in Rocky Mount, North Carolina, 1900–1935

Conor Harrison

Introduction

On 6 July 1916, a group of African American residents from the Crosstown neighbourhood of Rocky Mount, North Carolina (Figure 2.1), presented a petition to the Rocky Mount Board of Commissioners requesting electricity service. The electric utility in Rocky Mount was municipally owned and operated with the all-white Board of Commissioners possessing complete control over the day-to-day operations of the utility. Among the eight to sign the petition was William Sawyer, a bricklayer; Cato Garner, a labourer at the Atlantic Coast Line railroad shops; and William Howard, a labourer on a local farm. The wives of Garner and Howard also worked as a laundress and in the tobacco industry respectively. While Garner's two children attended school, two of Howard's children worked in tobacco while the youngest boy (aged 13) worked as an elevator operator at a hotel (Hill's Directory Co., 1920; U.S. Census, 1920).

This group of petitioners is indicative of the working class residents of Crosstown, one of the two working and middle-class African American neighbourhoods in Rocky Mount during the early 20th century. With multiple family members working, these households were able to cobble together enough money each month that a supposed luxury like electricity would be deemed affordable. When several members of a neighbourhood came together to petition for an electricity service on their block all that was needed was for the all-white Board to approve their petition and extend the service to their houses. While it is not clear from the official meeting minute notes what became of this petition, judging by the fact that subsequent requests and petitions for an electricity service from Crosstown were rejected by the Board it is unlikely the request was granted.

In Rocky Mount, like many other small towns with a municipally owned electric utility in the American South, possessing the ability to pay for electricity was not the only determinant in the spread of residential electricity service. Rather, the race of the recipient often played the deciding role in how, when, and where electricity service was spread. While the municipal ownership of infrastructure is at times held in opposition to private control, control of the Rocky Mount municipal electric utility by the ruling white elite in the town ensured that the benefits of electrification, and not just the electricity service, accrued almost exclusively to

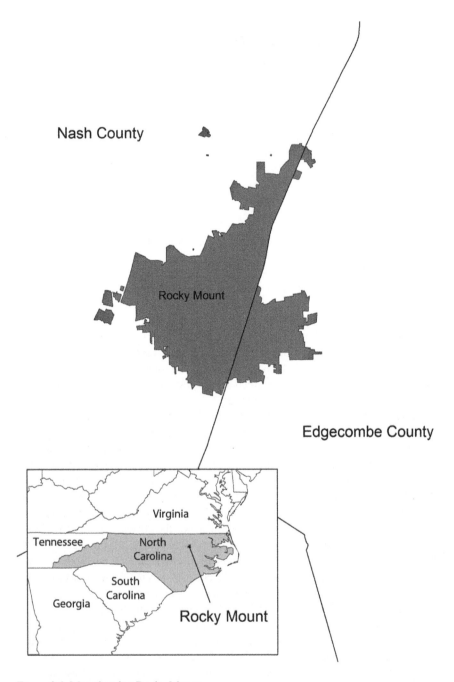

Figure 2.1 Map showing Rocky Mount
Source: the author

the white upper class of Rocky Mount. This has had long standing effects in Rocky Mount where today the town suffers from electricity prices 40 per cent higher than average in North Carolina. Much of this price differential is due to the operational practices established during the early years of the utility.

During the same period that the electricity service began in Rocky Mount, cities in the Southern United States were experiencing tremendous spatial and demographic change marked by rural to urban migration as well as a massive movement of freed slaves from rural areas into Southern cities. This shift occurred as subsistence farming withered due to shifts towards even greater dependence on cash crops like cotton and tobacco (Daniel, 1986; Rabinowitz, 1992). Displaced workers tended to drift towards small towns and cities where new industry was cropping up (e.g. iron in Alabama, coal in Appalachia, or textiles in the Carolinas). Manufacturing was drawn to the South by tax incentives as well as low wages and the lack of a unionised workforce. While larger cities like Atlanta and Charlotte attracted some manufacturing interests, these cities were increasingly focused on commerce and market functions. Rocky Mount's urbanisation, on the other hand, is similar to that of other 'second order' towns in the South which were primarily dependent upon a few manufacturing facilities (such as textiles and tobacco in Rocky Mount) drawing from local agricultural or natural resources (Cobb, 1988; Rabinowitz, 1992). Electric utilities were frequently spin-offs of these manufacturers' desire for industrial power. One such example, a hydroelectric venture of the American Tobacco Company in North Carolina, became Duke Energy, the largest electric utility in the United States in 2013.

The population of Rocky Mount, like many Southern towns, grew dramatically during the early 20th century, expanding from a small crossroads of several hundred people in 1880 into a fully-fledged town of 25,000 by 1930 (Hill's Directory Company, 1930). These new arrivals were met by new forms of spatial, social, and demographic control that were shaping growing cities in the South. Jim Crow laws, in combination with social codes and norms, legalised the separation of people based on race and provided one guiding planning ideology for the development of cities. This chapter shows that during the early parts of the 20th century, electricity service, far from being distributed equitably, was like other networked infrastructures employed as both a technology of rule and a producer of social and racial segregation and differentiation (Gandy, 2004; McFarlane, 2008; McFarlane & Rutherford, 2008; Connolly, 2009) by white elites in small Southern cities.

Simultaneously, Progressive Era ideologies of order and efficiency were dictating the reorganisation of cities into spaces of cleanliness, safety, and economic efficiency. Urban reformers during the Progressive Era believed that urban misery was not a permanent condition and as such sought ways to bring about an end to the crime, poverty, and poor health that had defined urban areas (Hofstadter, 1955; Wiebe, 1967; Grantham, 1983; Ayers, 1992). The belief was that by reorganising cities, clearing slum housing, and encouraging education and culture, many of the ills of city life could be eliminated. In most cities improvement was at least in part predicated upon the availability of networked urban infrastructures: water and sewage systems to alleviate water-borne disease; road and rail service to connect labour and employer in a growing city; and electric illumination and power in homes, stores, and factories (Tarr & Dupuy, 1988). Access to these technologies became a key component of constructing the city as a modern site of efficient capitalist production and expansion (Swyngedouw, 1999; Gandy, 2004; McFarlane & Rutherford, 2008). For many individuals and neighbourhoods, access to networked infrastructures was essential to success; denial, on the other hand, played a large part in reproducing inequalities, identities, and marking certain places and people as undeveloped and backwards (Graham & Marvin, 2001; Bakker, 2003; Kooy & Bakker, 2008; Zérah, 2008). But because electricity was not directly associated with public health, it is possible that it served as a more effective tool of segregating and differentiating Southern cities than other networked infrastructures.

Municipally owned and operated, the day-to-day operations of the utility were controlled by the Rocky Mount Board of Commissioners, whose meeting minutes were recorded in publicly accessible ledgers and later recounted in newspaper accounts. The availability of this data provides an opportunity to trace the growth of Rocky Mount as a city alongside the development of its electric utility, thus highlighting the co-evolving nature of cities and infrastructure (Coutard, 2008; McFarlane & Rutherford, 2008). Such a historical perspective provides new insights and perspectives into taken-for-granted historical accounts of the development of energy – principally an economic determinism used to explain the spread of electricity service and consumption. Tracing the development of electricity networks historically enables deeper consideration of the social, demographic, and political contexts in which it occurred (Hirsh & Jones, 2014), thus providing insights into the obduracy of particular networks (Hughes, 1987; Hommels, 2005), as well as the long-term uneven geographies they help to produce. Utility and municipal archives, like those drawn on for this chapter, provide evidence of the cultural and political dimensions, as well as the consequences, of the creation and use of energy infrastructures (Hirsh & Jones, 2014). In Rocky Mount and other cities, historical analysis shows that the actions of town elites can hardly be considered 'progressive'. Rather than progressive urban reform, electrification in the South was part of a project of social and demographic control and not solely the result of an economic geography of progress and efficiency.

How exactly did the municipal electric utility in Rocky Mount, and many other small Southern cities, serve to further the interests of the white elite? This chapter considers that question in two ways. It first examines the geography of requests for electricity service in the home between 1907 and 1923 by mapping the location of requests made before the Board of Commissioners. This shows that the Rocky Mount Board of Commissioners pursued a policy that deliberately steered electricity to white areas of town even when spreading electricity more broadly than would have been economically 'rational'. Second, it traces how the existence of a specifically municipally owned electric utility allowed for certain benefits to accrue to the town's white ruling class. An analysis of the changes powering industry, municipal budgeting, and electricity unaffordability provides evidence that this was the case. The conclusion points to the ways in which this work calls into question the economic determinism that shapes most histories of the dispersal of residential electricity use and posits that electricity may have proven a more effective, palatable, and economically beneficial means of segregating and differentiating towns than other networked infrastructures.

Jim Crow and the Progressive Era in North Carolina

The town of Rocky Mount today straddles the border of Edgecombe and Nash Counties in the largely rural northeastern North Carolina (Figure 2.1). Located on the western edge of the coastal plain, this region, and especially Edgecombe County, was home to some of the wealthiest antebellum planters in North Carolina and was one of only five counties in North Carolina that had more than 10,000 slaves (Weiler, 1991). Agriculture was exceedingly important to Edgecombe County: in 1860, it was the largest cotton producer in the state and cotton remained the driver of the local economy until the 1890s when a drop in cotton prices brought about a shift towards tobacco. In subsequent years farmers tended to switch their emphasis back and forth between cotton and tobacco depending on prevailing prices, but by the 1920s tobacco was king (Weiler, 1991).

For a brief period after the United States Civil War, the traditional ruling white planter class found themselves outnumbered by freed blacks and their ruling power circumscribed by Federal troops and Reconstruction laws. Local politics dramatically changed with numerous blacks holding local, state, and federal offices. This period was short lived as by 1870 the conservative Democratic Party had retaken control of the North Carolina state legislature and by 1875 state and county politics were largely back in the control of a minority of white landowners. A series of reconfigured county borders and new laws enabled the majority white state legislature to regain control over majority black counties like Edgecombe County (Weiler, 1991; Southern, 2005). Between 1880 and 1900, voting rights for blacks were gradually eroded via the selective implementation of literacy tests, landholding requirements, and poll taxes at the local level. In conjunction with the overall decline in black population in Edgecombe County (a decrease of nearly 3,500 people between 1880–1890), the more progressive Republican Party's grip on local politics began to decline. In the wake of the 1898 state elections, race riots sprung up across North Carolina and were violently put down by a mix of state militias and newly formed 'White Supremacy' clubs as well as a reinvigorated Ku Klux Klan. In 1925, North Carolina had an estimated 86 organised Klan groups with a total membership approaching 50,000. The North Carolina Klan also had friends in powerful places – North Carolina's Grand Dragon during the 1920s, Henry Grady, was a judge on the State Superior Court (Cunningham, 2012). Even as the formal Klan disbanded in the late 1920s, reports from Rocky Mount in 1930 show that a local Klan organisation gave financial support to a poor white family during the annual Christmas charity drive put on by the Salvation Army (Hazirjian, 2003). Klan membership, of course, was not a requirement for the practice of white supremacy, which became embedded in cities through the implementation of Jim Crow laws.

In the American South, developing systems of networked infrastructure in the early 1900s meant interaction with legal, political, and economic frameworks that were developed to institutionalise segregation and white supremacy (Woodward, 1955; Cole et al., 2012; Steedman, 2012). The laws and codes that governed the segregation of Southern cities were known as Jim Crow laws, a name derived from a popular entertainment show involving a white actor performing in blackface (Woodward, 1955). Jim Crow largely defined black social, cultural, economic, and political life, regulating such mundane actions as eating, drinking, working, schooling, churchgoing, and transportation (Rabinowitz, 1992). Jim Crow also came to shape where people lived in cities. While historically African Americans and whites lived near to each other, Jim Crow zoning laws began dictating where African Americans could live and how they moved around the city (Thomas & Ritzdorf, 1997). Despite laws that provision of public facilities was done on a separate but equal basis, in reality this was far from the case. In cities and rural areas, local elites worked to shape and alter service provision to meet their own political and economic goals (Badger, 2007). As this chapter shows, the result was a fragmented networked infrastructure more similar to Bakker's (2003) 'archipelago' than Graham and Marvin's (2001) 'modern infrastructure ideal'.

The same white elites implementing Jim Crow laws were spearheading the 20th century industrialisation of the South, commonly referred to as the 'New South' (Weiler, 1991; Beckel, 2011). While prominent Southern historian C. Van Woodward (1955) argued that the New South was largely controlled by a new class separate from old planter families, this view has more recently been refuted. Studies have shown that remnants of the planter class, through an alliance with northern capital eager to take advantage of the cheap labour and land in the South, have maintained dominance over African Americans and poor whites until present times (Rabinowitz, 1992; Woods, 1998). Further, because so few Confederate leaders actually lost land, nor the valuable social connections to important political institutions, a striking continuity of control in places such as Nash and Edgecombe County exists (Billings, 1979).

Such continuity is evident in Rocky Mount. Among the largest slaveholding families in the Rocky Mount area were the Battles, a family whose members would go on to achieve prominence in both local and state politics and exert strong control over the processes of urbanisation and infrastructure provision in Rocky Mount. For example, the nephew of William Battle, who owned 232 slaves before the Civil War and started the Rocky Mount Cotton Mills, was Thomas H. Battle. Thomas H. Battle possessed considerable social, political, and economic control over Rocky Mount through his ownership of Rocky Mount Mills (the largest manufacturing company in the region), two banks, and a real estate company. Battle would also serve ten years as mayor, more than 15 years on the Board of Commissioners, and 34 years as the chairman of the Board of Rocky Mount Schools (Weiler, 1991; Fleming, 2013). Thomas H. Battle could also draw on a myriad of important family and personal connections across the state, with his father Kemp serving as President of the University of North Carolina between 1876 and 1891 (Powell, 1996).

Thomas H. Battle was among the Rocky Mount Board of Commissioners that first started the municipal electric utility in 1902. The early 20th century in North Carolina was marked by 'business progressivism' – progressive policies that benefitted business – but there was little concern for social protections, particularly for African Americans. Progressivism generally was marked by a complex mix of social and political economic ideals, ranging from anti-corruption campaigns to prohibition to eugenics. A common thread among many Progressives was the link between the urban condition, race, and economic progress. Several leading economists, for example, argued that state aid to unfit workers – often the mentally ill, criminals, or so-called 'deficient races' – would provide conditions for these workers to easily multiply (Leonard, 2005). Social scientists such as Howard Odum of the University of North Carolina also employed crude hierarchies of race to explain away the 'Negro Problem' in their analyses of the American South (Woods, 1998). The near universal belief in white supremacy by the business and intellectual elites in the American South meant that the ideals of Progressivism and Jim Crow meshed together quite easily. In tandem, Progressivism and Jim Crow segregation operated as the central planning ideologies in many Southern cities during the early 20th century.

With a segregation-led division of races on the one hand, and a simultaneous quest for efficiency and order on the other, these ideologies combined to shape the emerging geographies of networked infrastructure in the South (Campbell, 1986; Wilson, 2002). Ideas of progress and efficiency were also shaped into more subtle methods for the control of races, what Connolly (2009) has referred to as 'technologies of racism'. These 'technologies' include many of the mundane and everyday practices involved in the governance of cities, including zoning, public transit, and building codes. What is important to recognise is that while these tactics were infused and shaped by racist ideologies, they also were employed to ensure conditions appropriate for capital accumulation (Marable, 1983; Connolly, 2009).

Historians of electricity, however, have been slow to incorporate ideas of social power and uneven development into their accounts of electrification. Most historians of electricity employ a sort of economic determinism that ascribes the spread of electricity to simple economic questions of rates, affordability, and consumer choice (Tobey, 1996). This view denies the inherently political nature of technologies and infrastructure (Feenberg, 1992; Graham & Marvin, 2001; McFarlane & Rutherford, 2008), as well as the fact that 'domestic electrical modernisation involved collective, public decisions' (Tobey, 1996: 3). Even further, many accounts ignore the 'interrelationship between materiality, governmentality, identity, and urbanisation' and 'how contested and evolving processes of social differentiation were linked to the differentiation of . . . infrastructures and urban spaces' (Kooy & Bakker, 2008: 1843). The following sections aim to begin filling this void by considering how the Rocky Mount electric utility was central to the ongoing and related processes of racial segregation and uneven economic development.

Residential Electricity Service in Rocky Mount

The first attempt to start an electric utility in Rocky Mount was by S. K. Fountain, who appeared before the Board of Commissioners in 1897 to request a franchise to operate an electric light plant and illuminate the town with arc lights. Fountain, a telegraph operator for the Atlantic Coast Line Railroad in Rocky Mount, was known as a local inventor and tinkerer who would later go on to start the first telephone service in the town. While the electric franchise was granted, no plant was ever built by Fountain (Rocky Mount Board of Commissioners [hereafter RMBOC], 6 July, 1897; Fleming, 2013). As evidence of the competition electric lighting faced as the lighting of choice in the late 19th century, several months later the town granted a franchise to Maurie Thomas and D. H. Whitehead to build an acetylene gas plant in the town to provide street lighting. A provision in the franchise, however, granted the city the right to buy the plant at any time for an agreed price (RMBOC, 2 February, 1898).

Not long after Thomas and Whitehead were granted their franchise, the town council began investigating forming a municipally owned electricity system on their own. To do so would require raising funds through the sale of municipal bonds, which at the time required the approval of the state legislature. Two prominent members of Rocky Mount, including the aforementioned Battle, were sent to the state capitol in Raleigh to try to secure passage of a bill allowing the bond issuance, which would then be put to a local vote for approval. The bonds would include funds for the electric light and sewerage system and the vote was cleverly tied into a bond issuance that would establish public graded schools in the town (Beck, 2002). After securing state legislative approval, a referendum was held on 19 March 1901. It secured overwhelming support and the town quickly began searching for a location to build the plant and a contractor to build it, as well as a bank to issue the bonds (Beck, 2002).

D. J. Rose, a member of the Board of Commissioners, was granted the contract to build the plant whose generator would be purchased from General Electric (RMBOC, 19 March, 1901). In June of 1901 construction began on the plant, with up to 50 convicts from the local prison hired to do much of the labour. This was a common practice in the South, with states frequently incarcerating blacks on trumped up charges in order to rent them out as labourers (Lichtenstein, 1996). By September of 1901, construction was going well enough that the plant superintendent was ordered to begin signing up customers and a scale of rates for the electric plant was adopted. The first street lights along Main Street were switched on in January 1902.

Mapping Electricity Service in Homes

Initially, the electric utility was primarily focused on providing street lighting for the town. By the 1910s, however, electricity service in private residences became more important. But before turning to an analysis of location of electricity service requests, it is important to take note of some key features in the urban space of Rocky Mount (Figure 2.2). The town is bisected by railroad tracks that also indicate the border between Nash and Edgecombe counties. The Edgecombe side (to the east) is, in general, considered the poorer side of the tracks. The area depicted as Church Street in the centre of the map is largely

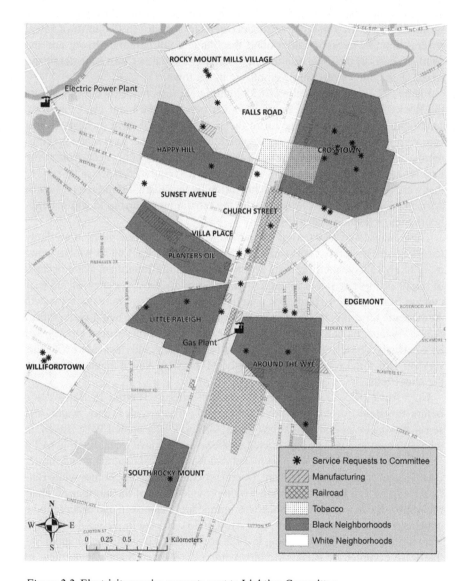

Figure 2.2 Electricity service requests sent to Lighting Committee

Source: the author, based on Hazirjian (2003); Meeting Minutes, RMBOC; Sanborn Fire Insurance Maps (1907, 1912, 1917)

comprised of a central business district located on both sides of the railroad tracks. To the northwest of town is the Tar River, along which the Power Plant was located after 1912, as well as the mill village of the Rocky Mount Mills (which was incorporated separately until 1928). To the south and southwest are two areas which were later annexed into the town: the white working class area of Willifordtown and the predominantly African American South Rocky Mount. The town's largest employers were mostly scattered on the Nash County side of the border, with a number of tobacco-related industry in the north of town, several manufacturers in and around the Planters Oil neighbourhood, and the southern parts of the town dominated by the large Atlantic Coast Line depot and repair shops, the largest single employer in the town.

Figure 2.2 shows the location of 35 lighting requests made by citizens before the Rocky Mount Board of Commissioners between 1907 and 1923 that were sent to the Lighting Committee for further deliberation. Of these requests, 57 per cent were made in African American neighbourhoods, 9 per cent made by commercial establishments, leaving slightly more than a third (34 per cent) from white neighbourhoods. Figure 2.3 shows an additional ten requests for electricity service that were granted without further deliberation by the Board. Of these two were in African American neighbourhoods with the remaining occurring in white neighbourhoods or commercial areas.

These maps show that during this period, African Americans were more likely to make requests before the Board of Commissioners for lighting, as well as being more likely to have their requests sent to the Lighting Committee. What became of many of these requests is unclear but judging by the frequency of requests coming from a similar geographic area like Crosstown, many were not approved. At the same time, the dearth of requests in the wealthier white neighbourhoods of Sunset Avenue, Falls Road, and Villa Place indicate that these areas already had an electricity service and that new homes there were readily provided with this service. This trend was evident in the case of the West Haven neighbourhood in the mid-1920s.

On 5 April 1923, D. J. Rose and the aforementioned Thomas H. Battle came before the Board of Commissioners requesting that the electricity, water, and sewer service be extended to a new neighbourhood, later named West Haven, in which 40 homes would be constructed on the western side of town (see Figure 2.4 for location). Approval of the plan meant that Rose (a local contractor and former Board member) would benefit from the construction of high end homes while Battle's Building and Loan association would provide many of the mortgages. Their request for the extension of service was immediately approved and ordered to be done as soon as possible (RMBOC, 5 April 1923). Homes in the neighbourhood were ultimately completed starting in 1928 and included large homes on large lots that were occupied by Rocky Mount's white elite (United States Department of the Interior, 2002).

As the following figure shows, the spread of electricity service in Rocky Mount was an uneven one shaped directly by racial discrimination. The effect of this was

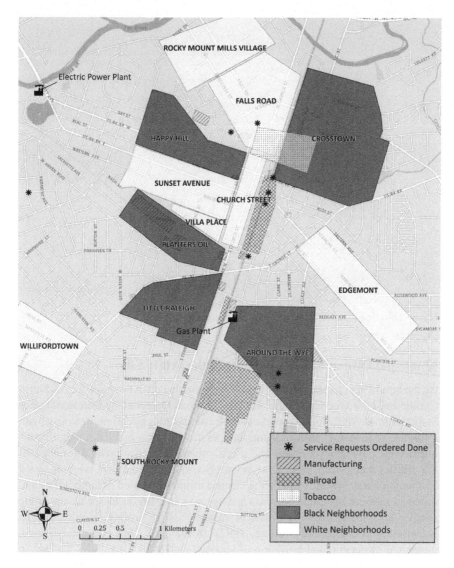

Figure 2.3 Electricity service requests ordered to be done by Board of Commissioners

Source: the author, based on Hazirjian (2003); Meeting Minutes, RMBOC; Sanborn Fire Insurance Maps (1907, 1912, 1917)

to mark certain areas of town as unmodern, dangerous, and unfit for receiving the privileges that come from full participation in democratic society. But controlling the operations of the municipal electric utility also provided further advantages to the Board of Commissioners.

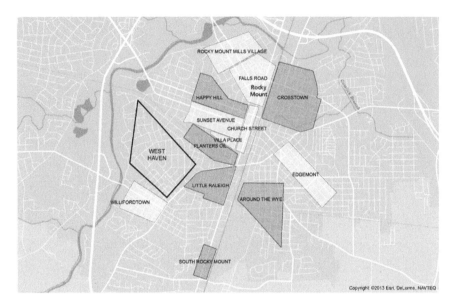

Figure 2.4 Location of West Haven neighbourhood
Source: the author

Municipal Ownership and Benefits to the White Elite

By the late 1890s electric utilities were being established in towns all across North
Carolina and the American South. The 1897 Annual Report of the Bureau of Labor
Statistics in North Carolina reported 22 electric utilities operating in towns, with
a further 75 'isolated' electric plants in operation, predominantly at manufacturing facilities, but also in hotels and a few private residences. The ownership of
the utilities was mixed, with some privately owned while others were operated
by municipalities. While most of the electric utilities operated in larger cities and
towns, a number of municipally owned systems had begun to spring up in the
smaller but rapidly growing towns of eastern North Carolina (North Carolina
Department of Labor and Printing, 1897). In the early 20th century, however,
municipal ownership and provision of services such as lighting and water was in
vogue among Progressive Era reformers, as between 1902 and 1930 the number of
municipally owned utilities actually increased in the United States (Schap, 1986),
and especially in small towns.

While the relative prevalence of municipal systems in smaller towns is often
attributed to their inability to attract private capital (Schap, 1986), as we have seen
in Rocky Mount attracting private capital was not likely to be difficult given the
personal wealth and connections of Thomas Battle and other Rocky Mount elites.
Some Progressive urban reformers in the late 19th and early 20th centuries were
eager to point out the advantages of municipal ownership in small towns, arguing
that, 'Here government lies close to the people. The officials are known to every

one. They cannot retire under the shield of their friends and party councillors. They are accessible to the personal complaints of every one' (Commons, 1904: 59–60). Municipal ownership of the electric utility, then, would be one part of 'cleaning up' towns, ridding them of corruption, and providing a cleaner, safer, and more healthful environment for its citizens. Critics are right to question these claims, however, because while there may be some difference in the degree of personal responsiveness in small communities, there was still ample opportunity for corruption, not to mention systematic institutional racism. As one critic points out, municipal ownership allowed for the appropriation of graft (i.e. kickbacks from construction or funnelling jobs to political supporters) in ways similar to the franchise system in large cities (Schap, 1986). But as MacKillop (2005) has shown, municipal ownership of electric utilities can provide indirect benefits to an elite ruling class – increasing the attractiveness of doing business in the community; increasing the length of the working day in their factories; and increasing the attractiveness of the community as a whole to stimulate population and business growth.

It is important to note that those who benefitted from electrification – both directly and indirectly – were local commercial elites, who in small Southern towns were often one and the same with local elected officials. So in effect, those who reaped the greatest benefits from municipally owned systems were those controlling municipal purse strings. Yet by financing these systems with municipal bonds, paying for the system was the responsibility of the community as whole. Municipal ownership of the electric utility was beneficial to the ruling elite in purely economic terms, but also in helping to order and segregate towns. The following considers three ways in which this occurred.

Electricity, Fixed Capital, and Industrial Benefits

Electric utilities require a tremendous upfront outlay of capital in order to begin operating. Because of this large outlay, and the relatively slow payback time on investment, the electric utility industry has long been dependent upon the state for its success (Howell, 2011; Harrison, 2013). In the United States, this came in the form of state regulation by the mid-1910s, but in the very earliest stages municipal ownership provided the most stable form of government intervention to ensure a utility's success. This took considerable pressure off of small factories eager to take advantage of electricity for lighting and motor driven machinery but unwilling to make their own investment in costly electricity generation equipment. With municipal ownership of the electric utility, an individual factory owner does not have their own capital tied up in fixed capital, and can use the electricity on a fee for service basis. This keeps more of their capital liquid, meaning it can be used more flexibly, for example invested in other ventures or used to expand production (Harvey, 2006).

Electricity provided multiple benefits to the factory owner. Before electricity, steam or waterpower were used to operate machines. In this particular arrangement, steam or water pressure powered a single turning mechanism, called the prime mover. The prime mover was then connected to an iron or steel line shaft that ran along the ceiling of the factory and connected to a series of leather belts and pulleys. A number of countershafts were then connected to the line shaft, and then via another set of pulleys and clutches ultimately to the machine. To operate a single machine meant activating a clutch to connect that machine to the vast web of belts and pulleys. Machines on different floors in a factory would be connected to the prime mover by belts running through holes in the floor, and these holes were then insulated to keep fire from spreading between floors (Schurr et al., 1990).

The provision of power via this extensive arrangement of belts and pulleys limited factories in two primary ways. First, an enormous amount of maintenance was required to keep the various moving parts in working order. Second, because of their connection to the line shaft, there was a limit to how machines could be organised, and they typically remained arranged in a linear fashion that was not necessarily ideal for production (Schurr et al., 1990). The arrival of electricity in the factory brought about rapid changes in this arrangement. By 1920 electricity was the predominant source of power in factories. Although electric motors were initially used only to turn the same line shafts, belts, and pulleys, they were soon used to power smaller segments of the factory, and ultimately individual machines. By 1929, the transition to electric drive motors was substantial, with 78 per cent of all mechanical drive in factories provided by electricity (Schurr et al., 1990). The widespread use of electricity in factories meant that redesigns of the factory floor, so influential to achieving assembly line efficiency under Taylorism, became possible. The precision of electricity – both in the ability to divide it into the most effective unit size and the ability to produce a motor to match the power produced exactly – helped drive an increase in factory throughput after World War One. Further, with the web of belts that drove the machines removed, overhead lighting from skylights and electric lights improved, making precision work easier (Schurr et al., 1990). In addition to the benefits electricity provided to the actual operations of the factory, using electric lighting in place of open flame lanterns or gas lighting decreased insurance rates due to the decreased risk of fires (Insurance Library Association of Boston, 1912).

In sum, electricity added flexibility and efficiency to the factory, making it very attractive to industrialists eager to take advantage of the latest methods of industrial organisation (Mumford, 1934). Increases in throughput meant that the cost of expensive machinery could be spread across the production of a greater number of more uniform units. However, these advances benefitted one class well above the rest: factory owners. Gains in assembly line efficiency, longer working hours, and more precision work all equated to the potential for increased profits among factory owners. In these ways, starting an electric utility, especially one that did not have their own personal capital invested in it, was of great benefit to the elite class of Rocky Mount and Southern industrialists.

Electricity and Municipal Budgets

In the early 20th century, many private electric utilities in larger towns were struggling to remain afloat. Because many cities granted multiple electricity franchises, the electric companies in those places engaged in fierce price competition that brought most to the brink of bankruptcy (Hughes, 1983). This was not the case with municipally owned utilities as the town itself was providing the service. With a captive potential customer base, especially among small to medium-size consumers unable to build isolated plants, the utility could charge the rates needed to at least break even. Most towns, however, sought to run their utility at a profit.

Rocky Mount was one such example. In 1902, its first year of operation, the electric utility turned a profit of $1,844, which was a profit margin of 55 per cent. In subsequent years this profit margin would fluctuate and at times be in the negative, especially in years with ambitious internally funded expansion programmes (as opposed to bond funding). After 1912, the profit margin would remain comfortably above 30 per cent (Table 2.1). Different than investor-owned electric utilities, these profits would not be retained for reinvestment in the business or distributed to shareholders. Instead, this money could be shifted around within municipal budgets to make up for shortfalls or cross-subsidising other municipal initiatives. Before 1910, the sum of electricity profits that could be shifted within the municipal budget was relatively minor, typically less than 5 per cent. That meant that a majority of municipal funding came from more traditional sources, namely property taxes and license fees.

By 1912 there was a dramatic shift as electricity profits began to routinely comprise more than 10 per cent of total revenue. In the mid-1920s, the town actually began to budget for a shortfall in general expenses, such as police, fire, and general operations, and plan to make up this shortfall with profits from the electric, and to a lesser extent, gas and water utilities. While this may appear a fairly mundane municipal accounting measure, it is important to consider what this practice made possible. First, in a city with a rapidly growing population desirous of ever-greater municipal services, property taxes could be held steady or even decreased. This was of overwhelming benefit to property holders in the town, who were predominantly white. Second, it was possible for the municipal government to present special tax incentives to attract new industries to town – offering a reprieve from property taxes for ten years, for example. Finally, by keeping electricity rates at a level high enough to maintain a healthy level of profitability, the benefits of electricity were by and large beyond the reach of the poor (this will be examined in more detail in the section that follows).

While rates were kept high enough to achieve healthy profits, the Board of Commissioners was cognisant of the effect high electricity prices had on industry and routinely acted to provide rate relief to industrial customers, even while requests for service from residential customers were being turned down. In one such example, in September of 1920 representatives from the Southern Cotton Oil factory and Ricks Hotel appeared at the Board of Commissioners meeting to request a decrease in rates. The representatives argued that the current rates were keeping business away from town and threatened to install isolated plants at their facilities if rates did not change. The issue was referred to the Light Committee to investigate (Rocky Mount Evening Telegram, 3 September, 1920a). Two weeks later, the Committee reported that they would hire an efficiency expert to examine

Table 2.1 Rocky Mount municipal revenues and profits, 1902–1929

Fiscal year	Profit	Profit margin	Profit as % of total municipal revenues
1902	$1,844	55.1%	7.3%
1903	$934	8.4%	4.2%
1904	$(7,507)	−61.5%	−17.9%
1905	$2,014	14.0%	4.1%
1906	$(2,209)	−14.0%	−3.8%
1907	$2,183	11.1%	3.1%
1908	$3,547	15.8%	3.7%
1909	$2,435	12.1%	2.9%
1910	$(7,254)	−32.5%	−8.7%
1911	$6,792	24.2%	7.9%
1912	$10,041	31.4%	10.9%
1913	$11,050	30.5%	10.8%
1914	$15,509	34.5%	9.9%
1915	$19,101	40.0%	9.8%
1916	$20,367	38.9%	11.8%
1917	$18,916	33.2%	9.1%
1918	N/A	N/A	N/A
1919	N/A	N/A	N/A
1920	N/A	N/A	N/A
1921	$28,625	18.1%	6.0%
1922	$63,117	43.5%	12.6%
1923	$63,478	35.4%	9.9%
1924	$77,792	37.8%	10.4%
1925	N/A	N/A	N/A
1926	N/A	N/A	N/A
1927	$128,150	51.9%	11.1%
1928	$130,159	52.0%	10.4%
1929	$142,608	53.5%	13.8%

Sources: author, from Meeting minutes of Rocky Mount Board of Commissioners; Rocky Mount Evening Telegram: Financial Audits of City of Rocky Mount

the possibility of decreasing rates, but in the meantime a larger discount would be applied to bills over $100/month (Rocky Mount Evening Telegram, 17 September, 1920b). By the middle of October, the Board reported that they would not hire an efficiency expert, but rather would offer increased discounts to those that used more electricity, a move that appeared to placate the protests of Southern Cotton Oil and Ricks Hotel (RMBOC, 11 October 1920).

The 'profits' from the municipal utility continued to be an important part of the budget of Rocky Mount well into the 1990s and have contributed to the problems of unaffordable electricity in the town today. But operating the utility at a profit was already creating problems for many customers by 1932.

1932 Electric Utility Charge-Offs

Rocky Mount was particularly hard hit by the Depression, with a jobless rate of almost 45 per cent by 1935. Once federal relief became available in the form of the New Deal Works Progress Administration, many Southern and Rocky Mount elites opposed the projects because they presented many poor people with an alternative to domestic work and intermittent and low-wage industrial labour (Hazirjian, 2003). After steadily increasing during the 1920s, the overall revenues for the electric utility had decreased slightly in 1932. Despite a decrease in revenues and hard times facing the city, the profit margin for the utility actually increased during the late 1920s. By essentially requiring the utility be profitable year on year, the municipal government was pricing electricity beyond the reach of many poor and working class customers, many of whom were undoubtedly facing challenges during the early years of the Depression.

Evidence of these challenges comes in the form of a list of 'charge-offs' from 1932. For electric utilities, a charge-off is a declaration that an account is unlikely to be collected and will be removed from the balance sheet and deducted from earnings. This printed list included 261 customers, the address at which service was received, and the balance on the account to be 'charged off'. Although they are not indicated as such, the charge-off list includes a mix of residential, commercial, and industrial customers. The balances left unpaid ranged from a few cents, like the $0.15 owed by Mozelle Davis (an African American cook living in the Happy Hill neighbourhood) to the $116.64 owed by Edgecombe Milling Company (located in the southeast of town along Cokey Road). Overall the median balance owed was $2.85. While a majority of the customers on the list appear to be residential, most of the unpaid money appears to be owed by industrial and commercial customers. However, there is also an uneven geography of where the customers with unpaid accounts are located.

Because home addresses are included in the charge-off list, the location of 257 of the 261 customer accounts can be accurately mapped. Of the total customers, 64 per cent (163) could reliably be coded as living in a black or white neighbourhood, with the remaining 36 per cent (94) located in the central business district or scattered in residential areas outside of the core neighbourhoods. As Figure 2.5 shows, 74 per cent (121) of the coded customers are located in African American neighbourhoods. Of these customers, the median account to be charged off is worth $3.21. Of the 26 per cent in white neighbourhoods, the median delinquent account value was nearly the same, about $3.15.

Table 2.2 shows how the location of charge-offs varied by neighbourhood. In general, African American neighbourhoods had the highest number of charge-offs, with the middle-class Crosstown neighbourhood leading the way with 53 charge-offs that had an average bill of $4.53. The white neighbourhoods typically have fewer charge-offs, but with a higher average bill. This is mostly due to the numerous businesses located in those neighbourhoods that skewed the average bill significantly higher. Mapping the uneven geography of electric utility

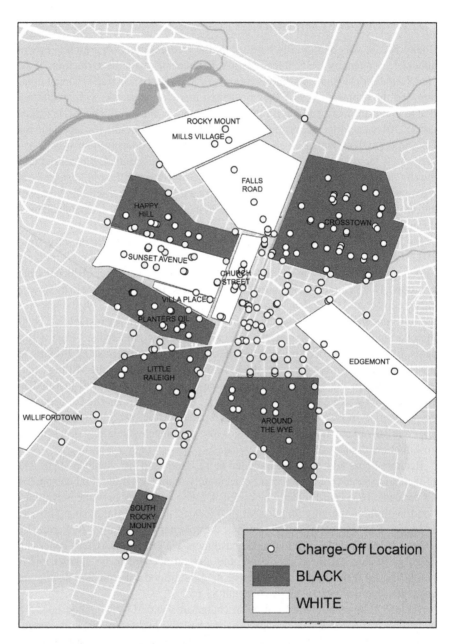

Figure 2.5 Location of charge-offs
Source: the author

Table 2.2 Charge-offs data by neighbourhood

Neighbourhood	No. charge-offs	Avg. bill
Crosstown	53	$4.53
Happy Hill	16	$5.55
Little Raleigh	15	$3.97
Around the Wye	17	$5.90
South Rocky Mount	3	$4.22
Planters Oil	17	$3.66
Willifordtown	0	N/A
Rocky Mount Mills Village	3	$5.76
Sunset Avenue	16	3.36
Villa Place	1	$1.38
Falls Road	5	$18.37
Church Street	14	$8.03
Edgemont	3	$9.69

Source: the author

charge-offs shows that African Americans were represented among the poor at far higher rates than whites. When compared to the maps of electricity service requests (Figures 2.2 and 2.3), it is evident that working class African Americans in neighbourhoods such as Crosstown and Happy Hill desired electricity, even going so far as to request it at the meetings of the Board of Commissioners. However, the persistent requests during the 1910s and 1920s show that electricity was slow to spread to those neighbourhoods. By 1932, the Board appears to have taken a more ambivalent position towards electricity service in African American neighbourhoods. Electricity use increased in African American neighbourhoods, as indicated by the number of delinquent accounts. But at the same time, the higher proportion of charge-offs in those areas points to two things. First, even middle and working class African Americans wages were so low that electricity was unaffordable. Second, and related to the first point, the insistence on the Board of Commissioners to run the electric utility at a profit continued to put electricity service outside the reach of the poor, and especially poor African Americans.

Conclusion

This chapter puts forward the case of electricity service in Rocky Mount as paradigmatic of the new tactics of social and economic control in the New South. Examining the growth of electricity service in the context of a growing Southern city provides insights into the ways in which urban infrastructures were key to the production of socio-spatial differentiation based on race. Close attention to the actual uneven diffusion of electricity service in Rocky Mount calls into question the economic determinism that most historians of electricity use to explain the growth of residential electricity demand. The historical approach is also of value to urban infrastructure studies in that it provides evidence of the cultural, political, and social origins of uneven service access. These patterns have long-lasting effects; even as the use of electricity in the home spread to African American neighbourhoods by the 1930s, the effects of running the utility in a way that benefitted white elites is quite clear.

Does electricity represent a unique case among networked urban infrastructures? The link between electricity and public health, unlike water, was not a direct one in the early 20th century. A lack of access to electricity, unlike clean water, did not directly pose sanitation or health issues to the broader population. As a result, municipal electric utilities could more easily be run at a healthy profit, and access could be denied without the spectre of disease and illness. This was certainly the case in Rocky Mount, where the municipal water system never amassed profits at anything near the rate as the electric utility (Figure 2.6). Because electricity was considered a household luxury until well into the 1930s, utilities were more focused on its use in industry and commerce (Nye, 1990). Because of this, it is possible that electricity access was a more effective and palatable technology of racism deployed by white elites in the American South.

Rocky Mount, and many other small towns in the American South, experienced revolutionary changes during the first three decades of the 20th century. The Progressive Movement brought about changes in the operation and governance of cities. Concerns over public health, corruption, and efficiency brought about large-scale restructuring of cities, especially in the new provision of the networked infrastructures of transport, water, gas, and electricity. The case of Rocky Mount

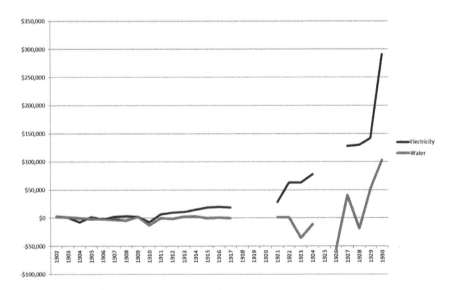

Figure 2.6 Rocky Mount electricity and water utility profits, 1902–1930

Source: the author, based on Rocky Mount Municipal Financial Statements, 1902–1930[1]

shows that the operation of the electric utility and the uneven distribution of the benefits of electricity by supposedly Progressive interests were predicated on a vision of white supremacy and racial antagonism that had long defined white-black relations in the South. In this way, a municipally owned electric utility appears ideally suited to the goals of both Progressivism and Jim Crow segregation.

Note

1 Data unavailable from 1918, 1919, 1920, and 1925.

References

Ayers, E.L., 1992. *The Promise of the New South: Life after Reconstruction.* New York: Oxford University Press.

Badger, A.J., 2007. *New Deal/New South: An Anthony J. Badger Reader.* Fayetteville, AR: University of Arkansas Press.

Bakker, K., 2003. 'Archipelagos and Networks: Urbanization and Water Privatization in the South'. *Geographical Journal* 169: 328–41.

Beck, B., 2002. *A Century of Service: City of Rocky Mount Public Utilities.* Rocky Mount, NC: City of Rocky Mount.

Beckel, D., 2011. *Radical Reform: Interracial Politics in Post-Emancipation North Carolina.* Charlottesville, VA: University of Virginia Press.

Billings, D., 1979. *Planters and the Making of a "New South": Class, Politics, and Development in North Carolina, 1865–1900.* Chapel Hill, NC: University of North Carolina Press.

Campbell, W.E., 1986. 'Profit, Prejudice, and Protest: Utility Competition and the Generation of Jim Crow Streetcars in Savannah, 1905–1907'. *The Georgia Historical Quarterly* 70(2): 197–231.

Cobb, J.C., 1988. *Industrialization and Southern Society, 1877–1984.* Chicago, IL: Dorsey Press.

Cole, S., N.J. Ring and P. Wallenstein (eds.), 2012. *The Folly of Jim Crow: Rethinking the Segregated.* South Arlington: University of Texas.

Commons, J.R., 1904. Municipal Electric Lighting. In: E.W. Bemis (ed.), *Municipal Monopolies.* New York: Thomas Y. Crowell, 55–182.

Connolly, N.D.B., 2009. 'Timely Innovations: Planes, Trains and the "Whites Only" Economy of a Pan-American City'. *Urban History* 36(2): 243–61.

Coutard, O., 2008. 'Placing Splintering Urbanism: Introduction'. *Geoforum* 39(6): 1815–20.

Cunningham, D., 2012. *Klansville, U.S.A.: The Rise and Fall of the Civil Rights-Era Ku Klux Klan.* New York: Oxford University Press.

Daniel, P., 1986. *Breaking the Land: The Transformation of Cotton, Tobacco, and Rice Cultures Since 1880.* Champaign, IL: University of Illinois Press.

Feenberg, A., 1992. 'Subversive Rationalization: Technology, Power and Democracy'. *Inquiry* 35(3/4): 301–22.

Fleming, M.S., 2013. *Legendary Locals of Edgecombe and Nash Counties, North Carolina.* Charleston, SC: Arcadia Publishing.

Gandy, M., 2004. 'Rethinking Urban Metabolism: Water, Space and the Modern City'. *City* 8: 363–79.

Graham, S. and S. Marvin, 2001. *Splintering Urbanism: Networked Infrastructures, Technological Mobilities and the Urban Condition.* New York: Routledge.

Grantham, D.W., 1983. *Southern Progressivism: The Reconciliation of Progress and Tradition.* Knoxville, TN: University of Tennessee Press.

Harrison, C., 2013. ' "Accomplished by Methods Which Are Indefensible": Electric Utilities, Finance, and the Natural Barriers to Accumulation'. *Geoforum* 49: 173–83.

Harvey, D., 2006. *The Limits to Capital.* London: Verso.

Hazirjian, L.G., 2003. *Negotiating Poverty: Economic Insecurity and the Politics of Working-Class Life in Rocky Mount, North Carolina, 1929–1969.* A Dissertation Submitted in Partial Fulfilment of the Requirements of Duke University for the Degree of Doctor of Philosophy. Durham, NC: Duke University.

Hill's Directory Company, 1920. *Hill's Rocky Mount, North Carolina City Directory 1920.* Richmond, VA: Hill Directory Co.

Hill's Directory Company, 1930. *Hill's Rocky Mount, North Carolina City Directory 1930.* Richmond, VA: Hill Directory Co.

Hirsh, R.F. and C.F. Jones, 2014. 'History's Contributions to Energy Research and Policy'. *Energy Research & Social Science* 1: 106–11.

Hofstadter, R., 1955. *The Age of Reform: From Bryan to F.D.R.* New York: Vintage Books.

Hommels, A., 2005. 'Studying Obduracy in the City: Toward a Productive Fusion between Technology Studies and Urban Studies'. *Science Technology Human Values* 30: 323–51.

Howell, J.P., 2011 'Powering "Progress": Regulation and the Development of Michigan's Electricity Landscape'. *Annals of the Association of American Geographers* 101(4): 962–70.

Hughes, T.P., 1983. *Networks of Power: Electrification in Western Society, 1880–1930.* Baltimore: Johns Hopkins University Press.

Hughes, T.P., 1987. The Evolution of Large Technological Systems. In: W.E. Bijker, T.P. Hughes and T.J. Pinch (eds.), *The Social Construction of Technological Systems.* Cambridge, MA: MIT Press, pp. 51–82.

Insurance Library Association of Boston, 1912. *Lectures on Fire Insurance: Being the Substance of Lectures Given Before the Evening Classes in Fire Insurance Conducted by the Insurance Library Association of Boston During the Fall and Winter of Nineteen Hundred and Eleven and Twelve.* Boston: Insurance Library Association of America.

Kooy, M. and K. Bakker, 2008. 'Technologies of Government: Constituting Subjectivities, Spaces, and Infrastructures in Colonial and Contemporary Jakarta'. *International Journal of Urban and Regional Research* 32(2): 375–91.

Leonard, T.C., 2005. 'Retrospectives: Eugenics and Economics in the Progressive Era'. *Journal of Economic Perspectives* 19(4): 207–24.

Lichtenstein, A.C., 1996. *Twice the Work of Free Labor: The Political Economy of Convict Labor in the New South.* New York: Verso.

MacKillop, F., 2005. 'The Los Angeles "Oligarchy" and the Governance of Water and Power Networks'. *Flux* 60–61: 23–34.

Marable, M., 1983. *How Capitalism Underdeveloped Black America.* Boston, MA: South End Press.

McFarlane, C., 2008. 'Governing the Contaminated City: Infrastructure and Sanitation in Colonial and Post-Colonial Bombay'. *International Journal of Urban and Regional Research* 32: 415–35.

McFarlane, C. and J. Rutherford, 2008. 'Political Infrastructures: Governing and Experiencing the Fabric of the City'. *International Journal of Urban and Regional Research* 32(2): 363–74.

Mumford, L., 1934. *Technics and Civilization.* New York: Harcourt, Brace and Company.

North Carolina Department of Labor and Printing, 1897. *Annual Report of the Bureau of Labor Statistics in North Carolina.* Raleigh, NC: Office of the Bureau of Labor Statistics.

Nye, D., 1990. *Electrifying America: Social Meanings of a New Technology, 1880–1940.* Cambridge, MA: MIT Press.

Powell, W.S., 1996. Battle, Kemp Plummer. In: W.S. Powell (ed.), *Dictionary of North Carolina Biography.* Chapel Hill, NC: University of North Carolina Press, pp. 114–15.

Rabinowitz, H.N., 1992. *The First New South.* Wheeling, IL: Harlan Davidson.

Rocky Mount Board of Commissioners, 1880–1930. *Rocky Mount Board of Commissioners Meeting Minutes.* Rocky Mount, NC: City of Rocky Mount.

Rocky Mount Evening Telegram, 1920a. Council Meeting News. *Rocky Mount Evening Telegram,* September 3.

Rocky Mount Evening Telegram, 1920b. Council Meeting News. *Rocky Mount Evening Telegram,* September 17.

Sanborn Map Company, 1907. *Rocky Mount, Nash and Edgecombe Co's, North Carolina, April 1907.* New York: Sanborn Map Company.

Sanborn Map Company, 1912. *Insurance Maps of Rocky Mount, Nash & Edgecombe Counties, North Carolina, July 1912.* New York: Sanborn Map Company.

Sanborn Map Company, 1917. *Insurance Maps of Rocky Mount, Nash & Edgecombe Counties, North Carolina, August 1917.* New York: Sanborn Map Company.

Schap, D., 1986. *Municipal Ownership in the Electric Utility Industry: A Centennial View.* New York: Praeger.

Schurr, S.H., C.C. Burwell, W.D. Devine and S. Sonenblum, 1990. *Electricity in the American Economy: Agent of Technological Progress.* New York: Greenwood Press.

Southern, D.W., 2005. *The Progressive Era and Race: Reaction and Reform, 1900–1917.* Wheeling, IL: Harlan Davidson.

Steedman, M.D., 2012. *Jim Crow Citizenship: Liberalism and the Southern Defense of Racial Hierarchy.* New York: Routledge.

Swyngedouw, E., 1999. 'Modernity and Hybridity: Nature, Regeneracionismo, and the Production of the Spanish Waterscape, 1890–1930'. *Annals of the Association of American Geographers* 89(3): 443–65.

Tarr, J.A. and G. Dupuy (eds.), 1988. *Technology and the Rise of the Networked City in Europe and America.* Philadelphia: Temple University Press.

Thomas, J.M. and M. Ritzdorf, 1997. *Urban Planning and the African American Community: In the Shadows.* Thousand Oaks, CA: Sage Publications.

Tobey, R.C., 1996. *Technology as Freedom: The New Deal and the Electrical Modernization of the American Home.* Berkeley, CA: University of California Press.

United States Department of the Interior, 2002. *National Register of Historic Places Registration Form – West Haven, Rocky Mount, NC.* Washington, DC: Government Printing Office.

U.S. Census, 1920. *United States Census.* Washington, DC: U.S. Government Printing Office.

Weiler, D., 1991. *An Examination of Political Hegemony: Race, Class, and Politics in Edgecombe County and Rocky Mount, North Carolina, 1865–1900.* Dissertation in Program in American Studies, Washington State University. Pullman, WA: Washington State University.

Wiebe, R.H., 1967. *The Search for Order, 1877–1920.* New York: Macmillan.

Wilson, B., 2002. 'Critically Understanding Race-Connected Practices: A Reading of W.E.B. DuBois and Richard Wright'. *The Professional Geographer* 54(1): 31–41.

Woods, C.A., 1998. *Development Arrested: The Blues and Plantation Power in the Mississippi Delta*. New York: Verso.

Woodward, C.V., 1955. *The Strange Career of Jim Crow*. New York: Oxford University Press, USA.

Zérah, M.-H., 2008. 'Splintering Urbanism in Mumbai: Contrasting Trends in a Multilayered Society'. *Geoforum* 39: 1922–32.

3   Plovdiv

(De-)racialising Electricity Access? Entanglements of the Material and the Discursive

Rosalina Babourkova

Introduction

Although technically covered by electricity distribution networks, residents of Romani neighbourhoods in Bulgaria have suffered from a differentiated utility service in the post-socialist period, both in terms of quality and quantity. This differentiated utility service reflects the everyday contradictions faced by Roma people in Bulgarian society, including the attitudes of non-Roma towards them. On the one hand, the stereotypical image of the Roma as the illiterate, 'backward other', living in abject poverty and lacking work discipline and respect for legal and social norms (Csepeli & Simon, 2004; Harper et al., 2009) fuels a variety of forms of discrimination across different domains of society, including service access. Such perceptions are reinforced by the widespread violation of building regulations characteristic of Romani neighbourhoods, which, although politically tolerated, prevents adequate legal access to urban infrastructure services (Babourkova, 2013). On the other hand, since the unbundling of the energy sector, pervasive non-payment and arrears in Romani neighbourhoods have led to increased disconnections, electricity rationing and discriminatory metering practices (Cohen, 2006; Zahariev & Jordanov, 2009; Babourkova, 2010). The accumulation of enormous individual and community debts in Romani neighbourhoods, encouraged by a recurrent 'infantilisation' and 'objectification' of the Roma by more powerful actors – from local politicians and 'human rights entrepreneurs' to European Union (EU) policy circles (Mcgarry, 2008, 2011; Trehan & Kóczé, 2009; Vermeersch, 2012) – can be traced to a history of political promises stating that the Roma do not have to pay for electricity given their disadvantaged condition. These factors have led to the development of a mostly subtle – yet often audible – discursive politics around race, ethnicity and access to electricity. Extreme right-wing parties stir racist sentiments against the Roma on the basis of their encroachment on basic services, especially electricity, with many ordinary Bulgarians siding with such views. Although not evident in the everyday interaction between the two ethnic groups, such views are strongly expressed in the media landscape and circulate across online forums.

The aim of this chapter is to demonstrate that the technical splintering of electricity services can serve to accentuate and augment existing divisions between ethnic or other social groups. Following an emergent literature on urban infrastructures and citizenship (McFarlane, 2008; McFarlane & Rutherford, 2008; Kooy &

Bakker, 2008a, 2008b; Anand, 2012; Rodgers & O'Neill, 2012; Wafer, 2012), the chapter examines the potential of seemingly neutral electricity networks to manifest racialised inequalities both in discursive and material terms. Drawing on an analysis of the discursive frames used in readers' comments to online news articles about the electricity rationing system in a Romani neighbourhood, the chapter unravels Bulgarians' moral discourses on entitlement and discrimination concerning access to electricity infrastructure services. By juxtaposing the racialised construction of two types of citizen-consumers – the 'undeserving Roma' versus the 'discriminated against Bulgarian' – a key message emerges: that the unequal geographies of the electric city are reproduced not only in the physical space of the city, but also discursively in the virtual (and electricity fuelled) worlds that urban citizens inhabit.

The story presented here is physically grounded in the Stolipinovo neighbourhood in the Eastern District of Plovdiv, Bulgaria's second-largest city (Figure 3.1).

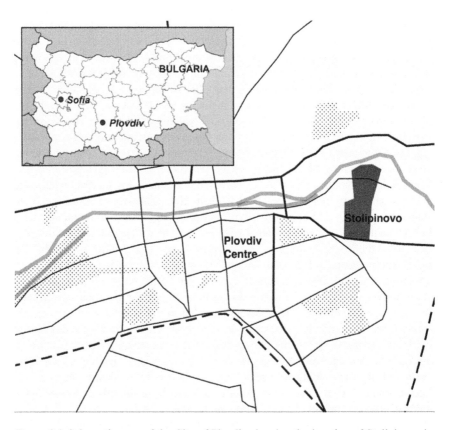

Figure 3.1 Schematic map of the City of Plovdiv showing the location of Stolipinovo in relation to the city centre

Source: the author, based on material prepared by Prathiwi Putri

Empirically, the chapter derives from an analysis of readers' comments to online news articles within the local press. Plovdiv boasts a population of around 340,000 people, with Turkish and Roma ethnic minorities officially making upto 5 per cent and 3 per cent of the city's overall population respectively (National Statistical Institute, 2012). Plovdiv's Roma population is concentrated in three ethnically segregated neighbourhoods, of which Stolipinovo with its estimated 40,000 inhabitants is among the largest so-called Romani 'ghettoes' in the entire Balkan region. Eighty per cent of the Roma in Stolipinovo live in 34 socialist-built condominium buildings, which combined boast 2,430 apartments. The rest live in self-built two-to-three-storey solid houses which lack building permits and proper access to networked infrastructure (Slaev, 2007).

Electricity provision in Stolipinovo has undergone significant changes and upheavals in the past three decades. Universal access to electricity, characteristic of the socialist period, continued through to the mid-1990s. Yet, following the accumulation of community debt through unpaid arrears – during the years of unbundling, corporatisation and privatisation of electricity distribution in the early 2000s – universal access was superseded by a near-withdrawal of the service in the neighbourhood. Manifested in a neighbourhood-wide electricity rationing regime from 2002 to 2009, this was not without social contestation (Cohen, 2006; Krastev, 2007). In recent years the Austrian energy company EVN, which acquired electricity distribution in south-eastern Bulgaria in 2005, has successfully upgraded Stolipinovo's electricity network and payment collection rates have soared. However, the general public's perceptions of the Roma as continuing to illegally encroach on the electricity network persist.

Race, Space and Oppression within Differentiated Access to Infrastructure Services

Graham and Marvin's (2001) splintering urbanism thesis provides a useful entry point for conceptualising the linkages between race, space and differentiated infrastructure provision. It charts and explains the rise of the 'modern infrastructure ideal' as the regulated and near universal access to infrastructure networks across western cities and its subsequent collapse which paved the way for a transition to privatised infrastructure service provision. The transition towards deregulation and privatisation of networked infrastructure services has proceeded unevenly in different parts of the world. In Central and Eastern Europe, for example, the liberalisation of the energy sector and privatisation of electricity distribution has framed the post-socialist geographies of energy transition (Bouzarovski, 2009; Bridge et al., 2013). In contrast, in cities of the global South, basic urban services have often remained in public (though more corporatised) hands, in order to serve local developmental agendas (Jaglin, 2008). Here, in some cases, the transition to privatised provision has even been reversed, particularly through the re-municipalisation of water supply systems (Pigeon et al., 2012). Regardless of the form of basic infrastructure provision, i.e. whether through public or private agents, infrastructural unbundling impacts on urban socio-spatial relations through the emergence of different forms of infrastructural 'bypass' and service differentiation (Graham & Marvin, 2001; Coutard, 2008).

The splintering urbanism thesis, as well as many of its critiques, locates the splintering/integration of urban infrastructure networks around the classic socio-logical category of class (Fernández-Maldonado, 2008; Pflieger & Matthieussent, 2008; Zérah, 2008). The differentiation in terms of network coverage and ser-vice quality is usually between poor versus rich areas of the city. However, class intersects with other identity dimensions and residents of poor areas can often be identified along racial, ethnic or religious lines as well. This is most obvious in the case of post-apartheid racial segregation in urban South Africa (Jaglin, 2008; Wafer, 2012). However, even in this context, racialised service differentiation between the former white (and hence well-off) city and the poor black townships is, to some extent, being overcome. For example, after the municipality of Cape Town ran a successful campaign exhorting all consumers to adopt pre-paid meters (PPMs), their discriminatory effects on the poor (black) man's meter appear to have subsided, at least to some commentators (Jaglin, 2008). The success of PPMs in Cape Town may exemplify a case where racial divides in energy infrastructure provision are being transcended. Yet, as such, Jaglin's analysis of basic service provision in Cape Town also continues along the class dimension even though it remains situated within longer racial histories.

Explicit constructions of difference and inequality along the lines of ethnic/racial identities rather than class are evident most clearly in historical analyses of urban infrastructure provision (McFarlane, 2008; Kooy & Bakker, 2008a, 2008b; see also Harrison, this volume). This is achieved through an analysis of the discursive and material shaping of urban infrastructure, thus addressing 'a moral urban politics based on the enrolment of subjects into "civilised" behav-iour' (McFarlane & Rutherford, 2008: 367). Kooy and Bakker (2008a) explain the fragmented nature of water supply interventions in colonial Jakarta as attempts to maintain and reinforce class and racial divisions between the city's residents. Similarly, Harrison (this volume) attributes to race the significant difficulties faced by black neighbourhoods in accessing electricity services in the early 20th century American South. In the case of Jakarta, the exclusive provision of both off-grid and networked water supply to 'modern' European populations divided the colonial city through a system of racialised modernity. This splintering from above was accompanied by a splintering from below, as 'residents whose identity as non-citizens precluded access to piped water (. . .) developed their own alterna-tives' (Kooy & Bakker, 2008a: 1855).

In the post-colonial era, however, water-related discursive metaphors of urban policy and planners have largely shifted from racial to socio-economic markers of status (Kooy & Bakker, 2008b). This shift points to an increasing difficulty in locating racialised constructions in infrastructure policy and planning. Construc-tions of race and ethnicity as drivers of unequal service provision may nowadays only be rendered visible through in-depth ethnographic research, such as Nikhil Anand's (2012) studies showing that Mumbai's most severe water problems – and unauthorised connections – are concentrated in predominantly Muslim areas.

Similarly to Kooy and Bakker's (2008a, 2008b) and McFarlane's (2008) historical discussions of racialised modernity, Anand's (2012) ethnography reveals how the water engineers of present-day Mumbai look upon residents of Muslim settlements as 'dirty' and 'not good', a contributing factor precluding them from legally accessing the municipal water service.

Anand (2012) explains the differential treatment of Muslim settlers through Julia Kristeva's notion of *abjection*. Political philosopher Iris Marion Young (1990) and geographer David Sibley (1995) also refer to Kristeva's theory of abjection to explain how the feeling of aversion within one group towards another lies at the root of cultural imperialism, operating as a form of oppression and putting in place a geography of exclusion. 'When the dominant group defines some groups as different, as the Other, the members of those groups are imprisoned in their bodies as the dominant discourse defines them in terms of bodily characteristics, and constructs their bodies as ugly, dirty, defiled, impure, contaminated, or sick' (Young, 1990: 124). Abjection therefore justifies particular forms of governing and the exclusion of social groups constructed as abject (Sibley, 1995), designating them to 'unlivable and uninhabitable zones of social life' (Murphy, 2006, cited in Anand, 2012: 488). For Anand, abjection constitutes, 'a dialectical process produced out of deeply situated discursive relationships and material practices, where difference is constantly reproduced, enacted and foregrounded between people that have deep overlapping social histories. Critically, these differences are realized and reproduced through the production and management of urban infrastructure' (Anand, 2012: 490).

Drawing on these scholars, this chapter argues that electric power grids, whether by (bad) fortune or design, can also manifest racialised inequalities, and that such racialised inequalities in energy access can be discursively reproduced through a language of abjection and hence oppression. Before turning to the methodological considerations and analysis of online commentaries to evidence of such racialised inequality, the next section provides a recent history of the nature of electricity access in Romani neighbourhoods in Bulgaria.

Network Splintering and (Re-)integration in Romani Neighbourhoods in Bulgaria

For the vast majority of urban Roma, just like for all Bulgarians, during late socialism electricity was a simple feature of everyday life – dirt-cheap, reliable and taken-for-granted (Krastev, 2007). Guided by the Leninist credo, 'Communism is Soviet power plus the electrification of the whole country', the energy sector in socialist states across Eastern Europe was developed to fully serve social needs and industrial development. All energy-related activities were centrally planned by the state, with the aim of ensuring universal energy access for domestic consumers whilst fuelling the growth of heavy industry (Bouzarovski, 2009). In post-war Bulgaria, the nationalisation of industry, the energy sector, the mines and the banks fuelled the country's power generation, distribution and supply. As the socialist General Development Plans for urban areas included dedicated sections on power engineering, urban electrification in the 1950s and 1960s neatly followed the development of the modern infrastructure ideal (Spirov, 2011a). By 1970 the electrification rate neared 100 per cent. At the end of the socialist period, the country's annual per capita electricity consumption was on par with industrialised European nations (Spirov, 2011b). Under state socialism, electricity tariffs were heavily subsidised, to the extent that it was nearly a free service. Romani neighbourhoods were successfully integrated into the socialist government's modern infrastructure ideal via electricity access, although they remained neglected in terms of other urban infrastructure services.

Over the course of the early transition towards democratic capitalism, whilst electricity distribution companies (EDCs) were still in public hands, successive governments across the spectrum (from socialist to right-wing) continued to use the electricity system as a form of unofficial social redistribution policy. This would mean a marked tolerance towards non-payment, despite a liberalisation package already agreed with the World Bank and the International Monetary Fund (Velody et al., 2003; Krastev, 2007). In parallel, as a strategy to gain votes, political parties embarked on the practice of promising free electricity to residents of Romani neighbourhoods (Babourkova, 2010). In 1998, for example, Spas Garnevski, the then mayor of Plovdiv and a prominent leader of the pro-democracy Union of Democratic Forces, publicly announced that 'the Roma are socially disadvantaged and therefore do not have to pay their electricity bills' (Cohen, 2006: 18). Statements like this rapidly fuelled a rise in non-payment and illegal connections to the grid. Despite the amount of negative coverage towards such practices within the local and national press, municipal and national governments as well as the publicly owned EDCs continued to ignore them. This political courting of Romani communities, also heavily covered by the national media, outraged many Bulgarian voters who felt ignored by the politicians and reinforced existing negative feelings towards the Roma (National Democratic Institute for International Affairs, 2006).

Throughout the 1990s and early 2000s, the publicly owned EDCs turned electricity provision for Romani neighbourhoods into a matter of ethnic politics. This politicisation was partly the result of public outrage against the state bestowing extra benefits on the 'undeserving' Roma. It was also derived from external pressure to reduce electricity theft rates and increase collection rates, in light of ensuing deregulation and corporatisation of the energy sector as a whole and of the privatisation of electricity distribution in particular. In the late 1990s electricity meters in Romani neighbourhoods began to be moved out of people's homes and locked up in electricity meter boxes between 6 and 12 metres in height – a practice described by activists as 'surrealist antirobbery experiments' (Zoon, 2001: 138). With their tall metering towers, Romani neighbourhoods became distinguishable from afar, identified as 'ghettoes' by the city's population (Figure 3.2). Placing electricity meters in inaccessible locations not only prevents consumers from checking their electricity meter readings, it also presumes that Roma people are more prone to illegally access infrastructure services, perpetuating established stereotypes of Roma people as only capable of deviant or criminal behaviour (Sibley, 1995; Bancroft, 2005; Trehan & Kóczé, 2009). Raising electricity meters and securing

Figure 3.2 Raised electricity meters in Stolipinovo before 2009
Source: Boyan Zahariev, Valeri Lekov and Ilko Jordanov (photo copyright)

them with alarms designates Romani neighbourhoods as problem areas, operating outside the norm and where the use of special measures – deviating from both law and custom – is unquestionably justified (Dikeç, 2001; Wacquant, 2007).

Another technical splintering measure of that time period particular to Stolipinovo was the imposition of an electricity-rationing regime upon the entire neighbourhood. Stolipinovo's main electricity supply problem had been the community's accumulated debt from non-payments of electricity bills since the early 1990s. After a sudden electricity cut in February 2002, the community took to the streets to protest resulting in damaged property and clashes with the police. Since this incident, throughout the majority of the 2000s, the entire neighbourhood only received electricity service from 7pm to 7am (Cohen, 2006). In the summer months, electricity supply was turned on only after 10pm. The majority of Stolipinovo residents reportedly used gas bottles for cooking because there was no electricity during the day. The lack of electricity also caused problems for food preparation and storage, as the intermittent electricity supply prevented the use of refrigerators (Krastev, 2007; Babourkova, 2010).

In 2005, after the Austrian utility provider EVN acquired *Elektrorazpredelenie Plovdiv*, local and national political figures, especially of the locally powerful Movement for Rights and Freedoms, attempted to negotiate an end to the rationing regime with the new owners. However, EVN initially refused to end electricity rationing for the neighbourhood (Terzieva, 2005). Following this, the Plovdiv-based NGO Association for European Integration and Human Rights made a formal complaint to the Commission for the Protection Against Discrimination on behalf of Stolipinovo resident Mehmed Denev. In addition, the NGO launched a court case on behalf of the same resident against the city's utility provider at the Plovdiv District Court. The Association for European Integration and Human Rights argued that the rationing regime affected both the non-paying and the regularly paying customers, such as Denev, with the latter facing a double disadvantage. *Elektrorazpredelenie Plovdiv* maintained that the reason for the rationing had not been (ethnic) discrimination, but the mass non-payment of electricity bills together with actions to stop or delay the work of bill collectors and the large volumes of unauthorised usage of electricity in the neighbourhood. Collection rates in Stolipinovo were estimated to be less than one per cent at the time. Out of 4,800 existing subscribers, only 254 had been regularly paying their bills. Of these, 54 were found to be unfairly suffering from the rationing system, as reported by the company (*Novinar*, 2006).

Denev's lawyers reported that, '. . . daily, "with the break of dawn", the electricity supply in his home is stopped. Only at night can Denev's family use the civilising benefits of electrical energy. Exceptions are not even made for weekends and public holidays' (Chernicherska, 2006). In October 2006 the Plovdiv District Court acknowledged the ethnically discriminatory character of electricity rationing in Stolipinovo, and ordered *Elektrorazpredelenie Plovdiv* to pay first 1,000 leva (€500) and later 3,000 leva (€1,500) in moral damages to Denev. The Court's decision emphasised that it is inadmissible for the EDC to treat paying customers in different ways: regularly paying customers in Stolipinovo, at the time affected by the rationing, should be treated in the same way as those who live elsewhere in Plovdiv (Chernicherska, 2006). This statement was particularly relevant since ethnically mixed and Bulgarian-only neighbourhoods, where non-payment also occurs, were spared such forms of collective punishment (Cohen, 2006; see also Zahariev & Jordanov, 2009). Probably as a result of these legal measures, EVN reported in 2007 that 19 of its regularly paying customers in Stolipinovo had been spared the electricity rationing continuing to afflict the rest of the neighbourhood (*Mediapool.bg*, 2007).

The situation in Stolipinovo reflected the challenges that foreign EDCs encountered after taking over distribution networks: high losses in transmission and distribution, paired with low revenue and cash collection. Whilst incoming foreign managers were baffled by the particularly high rates of electricity theft and non-payment in compact Romani neighbourhoods such as Stolipinovo (Coudenhove-Kalergi & Seelos, 2012a), it was acknowledged that both households and businesses carry out electricity theft, regardless of ethnic origin. Such practices are carried out with the help of corrupt staff of EDCs, leading to a surge in electricity theft rates particularly after electricity price hikes (Center for the Study of Democracy, 2010). After privatisation the government has been reluctant to get involved in disputes between consumers and suppliers around unauthorised usage of electricity, relegating all responsibility to find solutions to the problem to the private utility companies. Zahariev and Jordanov (2009) argue that the government's reluctance is reflected in the omission of non-payment and arrears issues in the 2002 Bulgarian Energy Strategy, which in turn reflects the state's neglect of wider issues affecting the country's poor (read 'Romani') neighbourhoods, such as the regulation of property rights, social and employment services and law and order.

Despite the lack of government support to solve non-payment and arrears issues, EVN embarked on a 3-million-leva (€1.5 million) upgrade programme in Stolipinovo in 2007. Over the course of the 2000s, Bulgaria's other two foreign-owned EDCs also invested in similar upgrade programmes, aimed at preventing theft and increasing collection rates in Romani neighbourhoods. Between 2007 and 2009 EVN upgraded transformers in Stolipinovo whilst replacing 178km of power cables and 5,500 electricity meters. EVN explained that precautionary measures were expected to improve collection rates (at the time less than 5 per cent), to be implemented alongside the settlement of debts, which in 2007 was estimated in 12 million leva (€6 million) (*News.bg*, 2007). EVN's Vice-President stated that regularly paying customers would have guaranteed supply, while non-paying account holders would see their supply terminated. The accumulated debts were to be settled through a system of deferred payments, meaning that subscribers would be offered to pay off a 'socially acceptable' sum of their accumulated debt alongside each regular monthly bill (*News.bg*, 2007).

Key to this re-integration and upgrade of Stolipinovo's electricity network and service has been EVN's approach to the local Roma community. Willing to work with local and national NGOs and to negotiate with the Roma and hear out their grievances, EVN adopted a new business logic counteracting traditional stereotypes towards the Roma. The company signalled equal customer treatment to Stolipinovo residents by lowering electricity meters to eye level (Figure 3.3), despite being advised against this measure by the regulator, in order to allow customers to track their consumption (Coudenhove-Kalergi & Seelos, 2012b). EVN ended the neighbourhood-wide electricity rationing regime, installed new individual meters for previously unconnected households and improved communication with the local community by employing local residents and opening up paying points in the neighbourhood aimed at increasing willingness to pay (*Sega*, 2013). Despite measures to create a positive relationship between the Roma community and the company (Coudenhove-Kalergi & Seelos, 2012b), the wider discursive effects of racialisation through the earlier political and material splintering of electricity distribution in Romani neighbourhoods continue to persist.

Figure 3.3 Lowered electricity meters in Stolipinovo after 2009
Source: the author

The Discursive Politics of Electricity Rationing in Plovdiv's Stolipinovo Neighbourhood

Discourses of Electricity Access in the Ghetto: Methodological Considerations

The overtly racist nature of comments posted by anonymous readers in response to most online news articles becomes acutely apparent when trawling the Bulgarian media landscape for issues of electricity access in Romani neighbourhoods. As Young (1990) argues, much of the aggressive experience of affectively 'othering' people different from oneself occurs in mundane contexts of interaction – in gestures, speech, tone of voice, movement and reaction of others. With the Internet offering unique ways of expressing the self and constructing social reality (Wittel, 2000; Sade-Beck, 2004), online commentaries have become one of these mundane contexts of interaction and voicing opinions. Online commentaries reflect an unmediated and unrestrained outburst of emotions amongst users. Situated in the context of a news item, online commentaries exhibit a linkage to the real, material world out there (in the form of the real-life event reported) as opposed to the often criticised and physically decontextualised environment of other forms of online interaction such as specialist forums, online chats and closed groups (Hine, 2000). To an extent, online commentaries become an extension of the article itself. Yet, probably because of the limited interaction between commentators, these commentaries stand as overlooked sites for research by critics and proponents of virtual ethnography (Wittel, 2000; Beaulieu, 2004; Sade-Beck, 2004; Garcia et al., 2009). While most commentaries can be taken out of the 'conversation' and examined as stand-alone personal opinions on the subject of the article, they are nonetheless valid responses and interpretations of the news topic, providing insights into how the commentator constitutes a reality around the topic.

Scary and intriguing at the same time, the commentaries analysed below are evidence of the brooding hostility towards the Roma on the part of the broader Bulgarian public. Establishing a contrast with inter-ethnic street interactions, where hostility is not necessarily manifest and open, these commentaries are evidence of the making of a racialised discourse around access to urban infra-structure. They render a new reality where the physical and spatial injustices of the sky-high electricity meters and other service differentiations in Romani neighbourhoods abandon the physical boundaries of the ghetto, yet enter a more subtle realm of (social) justice and the politics of difference (Young, 1990). The physical and spatial injustices that a minority group suffers become discursively regulated, i.e. reinforced and stabilised, through the everyday discourse of the majority. The online commentaries to specific news articles about the rationing system in Stolipinovo present clear and public sites of the intersection between inter-ethnic difference and issues related to infrastructure access. As such, the reality of oppression is first and foremost constructed by members of the public themselves, even though there is no way of knowing who they are, and only at a later stage co-constructed by the researcher through the selection of the commentaries presented and their interpretation.

The analysis that follows is based on readers' comments to three news items on the electricity rationing in Stolipinovo dating between 2005 and 2007, just before the completion of the above-mentioned network upgrade in the neighbour-hood. The comments show how some of the discursive frames used by members of the public construct a notion of ethnicised oppression suffered by Bulgarian consumers that has allowed continued electricity access in Romani neighbour-hoods despite non-payments. All readers' comments were accessed online in June 2013 and have been translated from Bulgarian to better convey their meaning. However, the specificity of punctuation has been retained in order to maintain the tone of voice of the commentator.

Discriminated Against Bulgarians and Undeserving Roma: Perceptions of Oppression in Access to Electricity

An initial analysis of readers' comments highlights the difference between the tone of voice of the reader (e.g. sarcastic, self-deprecating or hateful) and the actual content of the comment. Most comments are made by one stereotypical character, which I call 'the discriminated against Bulgarian'. The discriminated against Bulgarian speaks with different tones of voice. A common one is a voice of deep irony and sarcasm, which can be targeted towards different things – the disillusionment with a corrupt political system or the perception of double standards imposed on the Bulgarian public through the privatisation (and westernisation) of utility services. One commentator mockingly derides the Austrian company EVN for discriminating against the Roma by maintaining the electricity rationing system:

> *... Hello-o-o the Austrians!!! What are these inhumane actions?? Electricity rationing?? For Roma?? In which century are we?? Is that what they teach you in Austria?? And if someone complains to your minority rights organisations?? Are we to blame again?? So [the Austrians] came here and Bulgarianised!!*
>
> (Dr. Kosev, 2 June 2005, 01:27, *Sega*)

The commentator alludes to the plethora of domestic and international NGOs whose work focuses exclusively on Roma rights and inclusion. Despite the mocking and humorous tone, there is a sense that whatever is done to the Roma, the average Bulgarian will always be (unfairly) blamed at the end. Especially sarcastic is the statement that the Austrian managers have 'Bulgarianised', i.e. that they have quickly adopted 'wicked' Bulgarian culture and the established Bulgarian ways of dominating and oppressing minorities. Similar sarcasm is also evident in the following comment:

> *Plovdiv's energo[1] should have been given to the South Africans, but the Austrians seem also to be on the right track ;-).*
>
> (Dushko, 2 June 2005, 02:01, *Sega*)

The allusion to South Africa in Dushko's comment above evokes the notion of apartheid existing between Roma and non-Roma and suggests that a fair solution for electricity distribution in Romani neighbourhoods could be achieved only through an apartheid system, of which the rationing regime is an example. Dushko lauds the Austrians for maintaining the apartheid system of electricity rationing in Stolipinovo. As such the comment echoes Trehan and Kóczé's (2009) argument about the neo-colonial nature of relations between Roma and non-Roma.

The discriminated against Bulgarian can also sound insulted and extremely annoyed with a given situation and their words hint at their powerlessness against a non-transparent, unjust and corrupt system. Powerlessness, one of Young's (1990) five forms of oppression, and hence injustice, is the notion through which the discriminated against Bulgarian constructs the injustice experienced by him/her through the state's and the private sector's special attention to the Roma. The perception that the Roma are accorded special attention stems from Bulgarian and EU affirmative policies and actions (e.g. Decade for Roma Inclusion 2005–2015) and from the perceived generosity of utility companies in dealing with Romani consumers (see also Zahariev & Jordanov, 2009). As such, the antithesis of the discriminated against Bulgarian becomes the 'undeserving Roma', who is always privileged by a corrupt political system and favoured by politicians (Trehan & Kóczé, 2009). A similar rhetoric of 'rights without obligations' is also evident towards South African township residents in relation to a self-reconnection campaign as a form of protest (Wafer, 2012). The following example illustrates this rhetoric in relation to unfair electricity access in Bulgaria:

> *How come these ciganyori² from Stolipinovo have electricity at night, when they are not paying!!! If I don't pay, no one will leave it on for me at night, because I am Bulgarian. Why don't I sue the state in Strasbourg for being made to pay my electricity regularly. . . .*
>
> (zApl3zzz, 2 June 2005, 03:29, *Sega*)

In his bewilderment and confusion over the injustice, this commentator suggests suing the Bulgarian state at the European Court for Human Rights in Strasbourg. The mention of the state shows the widespread perception of the Roma as being protected by it and receiving more social assistance than other marginalised (Bulgarian) groups. The ironic tone of voice, however, shows the absurdity and impossibility of the suggestion, further reinforcing a perceived powerlessness against the established situation. Picking up on the 'Strasbourg' theme, the following commentator even suggests that at the heart of the issue of electricity access are basic questions of citizenship, as the rights of (ethnic) Bulgarian citizens are not protected even by national legislation:

> *Lets sue the whole gypsy class in Strasbourg. It's about time, I believe, if the Bulgarian legislation does not guarantee the rights of Bulgarians.*
>
> (BOOGIE MAN, 2 June 2005, 10:53, *Sega*)

Such comments suggest that these contributors believe there is a conspiracy behind the plight of the average Bulgarian that can only be unmasked through scapegoating the Roma, and that it is only possible to correct this through cardinal measures towards the entire Roma ethnos. Other posts demonstrate the confusion and perplexity of the average Bulgarian about what is perceived to be an unjust situation; unjust not least because Bulgarians believe that utility companies recuperate accumulated debts from Romani neighbourhoods through increasing the bills of their regularly paying Bulgarian customers:

> *And why do they have electricity at all, if they are not paying? Bulgarians have their electricity cut off until they pay. Where is the discrimination?! We Bulgarians are discriminated in this case because we pay for their day tariff. Under 1% have apparently paid, it would have been funny, if it wasn't so tragic unfortunately.*
>
> (???, 7 September 2006, 16:26, *Novinar*)

Why do they have ELECTRICITY when they don't pay? Who pays for the electricity of the non-paying 4246 subscribers, us?

(Pitasht, 7 September 2006, 16:51, *Novinar*)

I am discriminated because I pay for their electricity. . . . Those who don't work have electricity, and those who work – must not eat. . . .

(444, 7 September 2006, 17:47, *Novinar*)

The last comment makes reference to the Bulgarian proverb 'the one who does not work, must not eat'. However, it is sarcastically reworked to mean that in reality those not working and instead living off benefits (i.e. the Roma) receive electricity for free, whereas the hard-working Bulgarian must not, or cannot, eat because there is no money left after electricity bills have been paid. This exemplifies the perception that honest, hard-working and tax-paying Bulgarians do not receive any assistance to survive in the market economy, but those living off the state are additionally allowed extra benefits (e.g. free electricity) that normal people must pay for under market conditions. As a commentator by the name of 'moni' says, '*mangalite*[3] (are allowed) to suck the blood out of the honest tax payer' (moni, 16 November 2006, 03:53, *Sega*).

The sarcasm and language of abjection of the discriminated against Bulgarian can assume extreme proportions when he or she takes on the voice of the stereotypical 'Gypsy' as in the following comment:

Hello, what are you thinking cutting off my electricity? That's what I am used to and I like it, electricity – ha, let the whites pay for it. Dogan thinks about us and is going to arrange things for us again. I have 13 children, (I) drink and steal day and night, and now you dare switch off my electricity, so I am not able to see where I put stuff. . . a-a-a, c'mon, please. . . .

(Drug, 2 June 2005; 20:07, *Sega*)

The commentator alludes not only to several prevailing Roma stereotypes, i.e. having many children, drinking and thieving, but also to a form of political protection, which, in the eyes of many Bulgarians, is accorded to the Roma community. Referring to Ahmed Dogan, the head of the previously mentioned Movement for Rights and Freedoms, the commentator hints that elections may be coming up and that local politicians will once again make political promises to the Roma in return for votes. A shorter example of the same rhetoric is evident in the post by a 'Mango' – a common colloquial nickname for someone of Romani origin – on the topic of EVN's upgrade of Stolipinovo's electricity network: '*I don't care – I am not going to pay*' (Mango, 12 July, 2007, 13:51, *Mediapool.bg*). As above, the comment shows the discriminated against Bulgarian's belief that whatever efforts the company may undertake to upgrade the network and improve collection rates, they are bound to be futile – the Roma are not going to pay their electricity bills regardless. From the point of view of the discriminated against Bulgarian, the

Roma do not care about what goes on in wider society and their (bad) habits cannot ever be changed. The persistency of this particular perception of the Roma as incorrigible came to light when, at a 2013 round table 'Facts against myths – social distances in Bulgarian society', organised by the Open Society Institute Sofia, a board member of EVN Bulgaria announced that nearly 95 per cent of Stolipinovo residents now pay their bills. The statement was so intriguing that at least three daily newspapers and three well-known online news agencies ran the story with the following headlines:

95% of Roma in Stolipinovo pay for electricity to Bulgarians' disbelief
(Sega, 2013)

95% of Roma in Stolipinovo pay for their electricity, Bulgarians do not believe
(24chasa, 2013)

Unbelievable, but fact: 95% in Stolipinovo pay for their electricity
(Novini.bg, 2013)

Despite EVN's statements, such headlines show how a significant sector of Bulgarian society still does not believe that Roma people actually pay for electricity (*Sega,* 2013). Unfortunately, the perception of Roma people as continuously sponging on the rest of society through illegal access to services and as undeservingly receiving help and money through ethnicised integration policies by both donors and the state (Harper et al., 2009; Trehan & Kóczé, 2009) has been particularly persistent and has further splintered relations between the two ethnic groups.

Conclusion

This chapter examined the potential of differentiated electricity services to drive and maintain existing racialised inequalities. In the case of Plovdiv, to counteract the pervasive non-payment in Romani neighbourhoods, first public and later privatised electricity companies physically and visibly reinforced the splintering of ethnic neighbourhoods through the placement of electricity meters 6 to 12 metres above ground. In Stolipinovo in particular, this inequality was exacerbated through the establishment of a neighbourhood-wide rationing regime. Such measures have reinforced existing perceptions of Romani neighbourhoods as ghettoes, and perpetuated the stigmatisation of residents as being only capable of deviant behaviour. They have also reinforced the belief that, alongside an infantilising condition, such deviant behaviour demands special measures to tame and civilise it.

The chapter also shows how both urban electricity networks and urban areas in post-socialist Bulgaria have been discursively splintered into paying/honest (i.e. ethnically Bulgarian) and non-paying/thieving (i.e. Roma) enclaves. The non-paying and thieving enclaves are those of predominantly Romani neighbourhoods, where growth and densification is largely associated with the state's lack of planning control and its inability to solve the housing issue for a particular social group in the years of the post-socialist transition. Hence there is a perception of a fundamental social injustice when, under market conditions, one's neighbour – whose difference is already constructed as abject (cf. Sibley, 1995; Harper et al., 2009) – is allowed to encroach on a basic service (e.g. electricity) or allowed to continue to receive it without having to pay. The chapter examined these perceptions of injustice through the discursive frames of what I have termed here the 'discriminated against Bulgarian': an individual who derives a sense of powerlessness, mistreatment and bewilderment from actions that appear to favour the 'underserving Roma', regardless of whether this is objectively true or not.

Probably unsurprisingly, it has taken an 'outsider' to the social history of electricity in Stolipinovo – the Austrian company EVN – to transcend the abjectly racialised relations between Stolipinovo residents and the previous publicly owned electricity provider. But even after EVN's inclusive (but remarkably understudied) approach to electricity network upgrade in Stolipinovo, with associated 'soft' measures to tackle non-payment and thefts, the average Bulgarian remains extremely reluctant to be persuaded that the Roma actually pay their electricity bills. The analysis of readers' comments to online news items reveals how the increase in payment rates by residents of Stolipinovo is disconnected from (and perhaps irrelevant for) the racially charged perceptions expressed by the 'discriminated against Bulgarian'.

Methodologically, the analysis of online readers' comments potentially opens up new avenues for researching public perceptions and the configuration of social stereotypes. Their wide availability and easy accessibility renders them attractive for data collection and analysis purposes. Beyond this, within the complex configuration characteristic of the relationship between urban politics and infrastructures, these comments serve to further the institutionalisation of existing stereotypes in everyday social interaction and to discursively reinforce the racialised inequalities inherent, amongst other things, in the energy infrastructure of Romani neighbourhoods.

Notes

1 *Energo* is a colloqiual term for an EDC.
2 This is a derogatory term for the Roma, based on the term *tsigani* (gypsies), made up by this commentator.
3 Another derogatory term for people of Roma origin.

References

24chasa, 2013. 95% ot romite v Stolipinovo si plashtat toka, bulgarite ne vyarvat (95% of Roma in 'Stolipinovo' pay for electricity, Bulgarians do not believe). [Online]. 28 January. Available from: http://www.24chasa.bg/Article.asp?ArticleId=1742468 [Accessed: 14 June 2013].

Anand, N., 2012. 'Municipal Disconnect: On Abject Water and Its Urban Infrastructures'. *Ethnography* 13: 487–509.

Babourkova, R., 2010. 'The Environmental Justice Implications of Utility Privatisation: The Case of the Electricity Supply in Bulgaria's Roma Settlements'. *International Journal of Urban Sustainable Development* 2: 24–44.

Babourkova, R., 2013. Informal Housing Production and Energy Poverty: The Law and Access to Electricity in a Romani Settlement in Sofia. Paper Presented at *Energy Vulnerability Conditions and Pathways Symposium*, MERCi, Manchester, 21–23 May 2013.

Bancroft, A., 2005. *Roma and Gypsy-Travellers in Europe: Modernity, Race, Space and Exclusion*. Aldershot: Ashgate.

Beaulieu, A., 2004. 'Mediating Ethnography: Objectivity and the Making of Ethnographies of the Internet'. *Social Epistemology* 18(2–3): 139–63.

Bouzarovski, S., 2009. 'East-Central Europe's Changing Energy Landscapes: A Place for Geography'. *Area* 41(4): 452–63.

Bridge, G., S. Bouzarovski, M. Bradshaw and N. Eyre, 2013. 'Geographies of Energy Transition: Space, Place and the Low-carbon Economy'. *Energy Policy* 53: 331–40.

Center for the Study of Democracy, 2010. *The Energy Sector in Bulgaria: Major Governance Issues*. [Online]. Sofia: Center for the Study of Democracy. Available from: http://www.csd.bg/artShow.php?id=15199 [Accessed: 1 April 2014].

Chernicherska, G., 2006. Rezhimat na toka v Stolipinovo – iztochnik na diskriminacia i korupcia (Electricity rationing in Stolipinovo – source of discrimination and corruption). *Mediapool.bg*. [Online]. 8 November. Available from: http://www.mediapool.bg/режимът-на-тока-в-"столипиново"-–-източник-на-дискриминация-и-корупция-news123182.html [Accessed: 14 June 2013].

Cohen, E., 2006. 'Stolipinovo Lives on Night Tariff from One Election to Another'. *Obektiv* 131: 18–19.

Coudenhove-Kalergi, B. and C. Seelos, 2012a. *EVN in Bulgaria (B) – Engaging the Roma Community*. [Online]. Stanford Center on Philanthropy and Civil Society. Available from: http://pacscenter.stanford.edu/sites/all/files/EVN%20in%20Bulgaria%20B_Engaging%20with%20the%20Roma%20community_2012.pdf [Accessed: 11 February 2015].

Coudenhove-Kalergi, B. and C. Seelos, 2012b. *EVN in Bulgaria (C) – Making it work*. [Online]. Stanford Center on Philanthropy and Civil Society. Available from: http://pacscenter.stanford.edu/sites/all/files/EVN%20in%20Bulgaria%20C_Making%20it%20work.pdf [Accessed: 11 February 2015].

Coutard, O., 2008. 'Placing Splintering Urbanism: Introduction'. *Geoforum* 39(6): 1815–20.

Csepeli, G. and D. Simon, 2004. 'Construction of Roma Identity in Eastern and Central Europe: Perception and Self-identification'. *Journal of Ethnic and Migration Studies* 30(1): 129–50.

Dikeç, M., 2001. 'Justice and the Spatial Imagination'. *Environment and Planning A* 33(10): 1785–805.

Fernández-Maldonado, A.M., 2008. 'Expanding Networks for the Urban Poor: Water and Telecommunications Services in Lima, Peru'. *Geoforum* 39(6): 1884–96.

Garcia, A.C., A.I. Standlee, J. Bechkoff and Y. Cui, 2009. 'Ethnographic Approaches to the Internet and Computer-Mediated Communication'. *Journal of Contemporary Ethnography* 38(1): 52–84.

Graham, S. and S. Marvin, 2001. *Splintering Urbanism: Networked Infrastructures, Technological Mobilities and the Urban Condition*. London and New York: Routledge.

Harper, K., T. Steger and R. Filcák, 2009. 'Environmental Justice and Roma Communities in Central and Eastern Europe'. *Environmental Policy and Governance* 19(4): 251–68.

Hine, C., 2000. *Virtual Ethnography*. London: SAGE.

Jaglin, S., 2008. 'Differentiated Network Services in Cape Town: Echoes of Splintering Urbanism?'. *Geoforum* 39: 897–1906.

Kooy, M. and K. Bakker, 2008a. 'Splintered Networks: The Colonial and Contemporary Waters of Jakarta'. *Geoforum* 39(6): 1843–58.

Kooy, M. and K. Bakker, 2008b, 'Technologies of Government: Constituting Subjectivities, Spaces, and Infrastructures in Colonial and Contemporary Jakarta'. *International Journal of Urban and Regional Research* 32(2): 375–91.

Krastev, S., 2007. *Power On, Power off. Access to Electricity and Roma in Bulgaria: From Universal Entitlement to Insecure Resource*. A Thesis Submitted in Partial Fulfilment of the Requirements of Central European University for the Degree Master of Arts. Budapest: Central European University.

McFarlane, C., 2008. 'Governing the Contaminated City: Infrastructure and Sanitation in Colonial and Post-colonial Bombay'. *International Journal of Urban and Regional Research* 32(2): 415–35.

McFarlane, C. and J. Rutherford, 2008. 'Political Infrastructures: Governing and Experiencing the Fabric of the City'. *International Journal of Urban and Regional Research* 32(2): 363–74.

McGarry, A., 2008. 'Ethnic Group Identity and the Roma Social Movement: Transnational Organizing Structures of Representation'. *Nationalities Papers* 36(6): 449–70.

McGarry, A., 2011. 'The Dilemma of the European Union's Roma Policy'. *Critical Social Policy* 32(1): 126–36.

Mediapool.bg, 2007. Zapochnalo e iznasyaneto na elektromerite v Stolipinovo (Moving out of electricity meters has begun in Stolipinovo). [Online]. 11 July. Available from: http://www.mediapool.bg/започнало-е-изнасянето-на-електромерите-в-столипиново-news130207.html [Accessed: 14 June 2013].

National Democratic Institute for International Affairs, 2006. *Roma Participation in the 2005 Bulgarian Parliamentary Elections*. [Online]. 18 January 2006. Washington, DC: NDI. Available from: https://www.ndi.org/files/1976_bg_roma_011806.pdf [Accessed: 31 July 2014].

National Statistical Institute, 2012. *National Register of Populated Places*. [Online]. Sofia: NSI. Available from: http://www.nsi.bg/nrnm/index.php?ezik=en&f=8&s=1&date=03.09.2013&e=1&s1=4&c1=1&a1=100000&c=0 [Accessed: 31 July 2014].

News.bg, 2007. 'Stolipinovo' bez rezhim na toka (Stolipinovo without electricity rationing). [Online]. 10 August. Available from: http://news.ibox.bg/news/id_1314379314 [Accessed: 14 June 2013].

Novinar, 2006. Stolipinovo: rezhimat na toka e diskriminacia (Stolipinovo: Electricity rationing is discrimination). [Online]. 8 September. Available from: http://novinar.bg/news/stolipinovo-rezhimat-na-toka-e-diskriminatciia_MjAyNzs3.html?qstr=столипиново [Accessed: 14 June 2013].

Novini.bg, 2013. Neveroyatno no fakt: 95% v Stolipinovo si plashtat toka (Unbelievable, but fact: 95% in Stolipinovo pay for their electricity). [Online]. 28 January. Available from: http://www.novini.bg/news/115039-невероятно-но-факт-95-в-столипиново-си-плащат-тока.html [Accessed: 22 July 2013].

Pflieger, G. and S. Matthieussent, 2008. 'Water and Power in Santiago de Chile: Socio-Spatial Segregation through Network Integration'. *Geoforum* 39: 1907–21.

Pigeon, M., D.A. McDonald, O. Hoedeman and S. Kishimoto (eds.), 2012. *Remunicipalisation: Putting Water Back into Public Hands*. [Online]. Amsterdam: Transnational Institute. Available from: http://www.tni.org/sites/www.tni.org/files/download/remunicipalisation_book_final_for_web_0.pdf [Accessed: 11 February 2015].

Rodgers, D. and B. O'Neill, 2012. 'Infrastructural Violence: Introduction to the Special Issue'. *Ethnography* 13(4): 401–12.

Sade-Beck, L., 2004. 'Internet Ethnography: Online and Offline'. *International Journal of Qualitative Methods* 3(2): 45–51.

Sega, 2013. 95% ot romite v Stolipinovo veche si plashtat toka za pochuda na bulgarite (95% of Roma in Stolipinovo already paying for electricity to Bulgarians' disbelief). [Online]. 28 January. Available from: http://www.segabg.com/article.php?id=634196 [Accessed: 14 June 2013].

Sibley, D., 1995. *Geographies of Exclusion: Society and Difference in the West.* London: Routledge.

Slaev, A.D., 2007. 'Bulgarian Policies towards the Roma Housing Problem and Roma Squatter Settlements'. *International Journal of Housing Policy* 7(1): 63–84.

Spirov, M., 2011a. 110 godini elektrifikacija na Sofia (110 year electrification of Sofia). *Publics.* [Online]. 10 February 2011. Available from: http://www.publics.bg/bg/publications/54/ [Accessed: 16 June 2013].

Spirov, M., 2011b. Treto pokolenie bulgarski elektroinjineri (Third generation of Bulgarian electrical engineers). *Publics.* [Online]. 6 April 2011. Available from: http://www.publics.bg/bg/publications/34/ [Accessed: 17 June 2013].

Terzieva, E., 2005. Avstriicite ostavyat rezhima na toka v Stolipinovo (The Austrians maintain electricity rationing in Stolipinovo). [Online]. 2 June. Available from: http://www.segabg.com/article.php?sid=2005060200010020002 [Accessed: 14 June 2013].

Trehan, N. and A. Kóczé, 2009. Racism, (Neo-)colonialism and Social Justice: The Struggle for the Soul of the Romani Movement in Post-socialist Europe. In: G. Huggan and I. Law (eds.), *Racism, Postcolonialism, Europe.* Liverpool: Liverpool University Press. pp. 50–73.

Velody, M., M.J.G. Cain, and M. Philips, 2003. *A Regional Review of Social Safety Net Approaches in Support of Energy Sector Reform.* Report Prepared for U.S. Agency for International Development (USAID), Europe & Eurasia/Economic Growth/Energy and Infrastructure Division (E&E/EG/EI). [Online]. Available from: http://pdf.usaid.gov/pdf_docs/PNACX443.pdf [Accessed: 27 July 2014].

Vermeersch, P., 2012. 'Reframing the Roma: EU Initiatives and the Politics of Reinterpretation'. *Journal of Ethnic and Migration Studies* 38(8): 1195–212.

Wacquant, L., 2007. 'Territorial Stigmatization in the Age of Advanced Marginality'. *Thesis Eleven* 91(1): 66–77.

Wafer, A., 2012. 'Discourses of Infrastructure and Citizenship in Post-Apartheid Soweto'. *Urban Forum* 23: 233–43.

Wittel, A., 2000. Ethnography on the Move: From Field to Net to Internet. *Forum Qualitative Sozialforschung/Forum: Qualitative Social Research.* [Online]. 1(1). Art. 21. Available from: http://www.qualitative-research.net/index.php/fqs/article/view/1131/2517 [Accessed: 23 April 2014].

Young, I.M., 1990. *Justice and the Politics of Difference.* Princeton, NJ: Princeton University Press.

Zahariev, B. and I. Jordanov, 2009. Geography of Exclusion, Space for Inclusion: Nonpayment of Electricity Bills in Roma Neighbourhoods in Bulgaria. In: K. Pallai (ed.), *Who Decides? Development, Planning, Services and Vulnerable People.* Budapest: Open Society Institute-Budapest. pp. 55–121.

Zérah, M.-H., 2008. 'Splintering Urbanism in Mumbai: Contrasting Trends in a Multilayered Society'. *Geoforum* 39: 1922–32.

Zoon, I., 2001. *On the Margins: Roma and Public Services in Romania, Bulgaria and Macedonia.* New York: Open Society Institute.

Part II
Rewiring the Urban Grid

4 Rio de Janeiro

Regularising *Favelas*, Energy Consumption and the Making of Consumers into Customers

Francesca Pilo'

Introduction

The process of regularising access to electricity services in informal settlements in cities of the South is often described as the 'transformation of consumers into customers' (USAID, 2009). This transformation consists, in most cases, of redefining a failing commercial relationship and standardising the behaviour of consumer households so that they adjust their power consumption to their perceived financial capacity. In the city of Rio de Janeiro, where almost 20 per cent of the population lives in informal settlements known as *favelas* (1,400,000 people according to the 2010 national census), the stakes are high given that electricity access is largely achieved via informal and individual (illegal) connections. Since the 1990s, the Brazilian restructuring and privatisation of the electricity sector and the concession of distribution systems to private operators has contributed to redefining how informal electricity access is managed. Arguably, the primary aim has been to recover the high commercial losses suffered by utilities. Although this underscores primarily an economic concern, it is, in fact, an eminently political intervention in the city.

The changing political landscape of Rio de Janeiro constantly reshapes the contours of the political and governance dimensions of the regularisation of electricity access in *favelas*. Given the control of a large number of *favelas* by drug factions – a characteristic of Rio since the 1980s – over the last five years *favelas* have become key sites of government action. This has been achieved through a new state public security programme that permanently installs police forces in some *favelas* through Pacifying Police Units (UPP in Portuguese): thirty-eight such units had been installed by 2014. This process has established a new context for electricity distribution companies in the city: a secure environment where the regularisation of services appears possible.

In the context of these major political transformations, this chapter explores the practical process of transforming electricity consumers into customers, with a primary focus on its energy-saving aspects. The regularisation of electricity services, beyond a strictly technical or commercial process (focused around either meter installation and network repair, or bill collection systems), involves specific tools to promote energy efficiency – with particular attention to user consumption patterns. The primary aim of the chapter is to show how this form of infrastructural reconfiguration is implemented through market logics, which in turn has consequences for the way in which energy-saving measures are promoted. The chapter unpacks how infrastructure is mobilised within the context of an energy efficiency policy and in an informal settlement context where market logics can be difficult to implement. For this purpose, the first section articulates two different bodies of literature that enable an understanding of the relationship between infrastructure, consumption practices and energy efficiency public policies.

The second section provides a short narrative describing the varied policies aimed at managing access to the electricity network in Rio's *favelas*. This historical overview is followed by the description of the regulatory context, the objectives and the implementation of the energy efficiency programme in *favelas*. The chapter closes with an analysis of the socio-technical tools used by the electricity utility company in the *favela* of Santa Marta to push households to save energy. This is based on fieldwork conducted in 2010–2011, based on semi-structured interviews with 25 households and with staff of the electricity distribution company, in particular those responsible for the energy-saving programme. This final section, by combining analyses of the tools implemented and of the perceptions of households, explores the limits of a market-led approach to energy saving and the contradictions of this policy. The chapter argues that the measures adopted by the electricity distribution company appear to both weaken the users' understanding of their own consumption patterns and also, paradoxically, to stimulate increased levels of consumption.

Connecting Political Infrastructures and Consumption Practices

Urban infrastructure studies have explored the role of these socio-technical systems in the fabric of the city from a wide variety of theoretical perspectives (McFarlane & Rutherford, 2008). A significant part of this body of literature has focused on the wave of privatisation in infrastructure services that began in the 1990s, with analysis centring on the resulting transformations in cities in both northern and southern hemispheres (Graham & Marvin, 2001; Jaglin, 2005; Coutard, 2008). In the case of cities in the global South, the term 'archipelagos', coined by Karen Bakker (2003), has been widely used to describe the multiplicity of forms of service provision encountered. Bakker contrasts the notion of archipelagos with the term 'network', arguing that the former better represents the urban infrastructural landscape of the global South. While Bakker talks about the presence of a range of 'off-network' services, the archipelago metaphor proves helpful in pointing to spatial disparities in service provision, particularly useful in describing conditions of access in Brazilian *favelas*. Largely informal, the network-related fragmentation is produced by the prior socio-political fragmentation of the urban fabric (Lopes de Souza, 2005): the territorial control over the *favelas* exercised by groups linked with drug trafficking.

Literature on infrastructure reconfiguration provides useful theoretical and analytical frameworks for analysing infrastructures as political tools capable of modifying power relations, of producing (or reducing) inequalities and of 'holding together' – not to say governing – the city (Lorrain, 2011). Historical analysis on water provision in Jakarta (Kooy & Bakker, 2008) and sewage infrastructures in Mumbai (McFarlane, 2008) show how, by generating notions associated with ideals of modernity, morality, public space and citizenship, infrastructure development ultimately comes to be an instrument for the construction of social inequalities (McFarlane & Rutherford, 2008). In her analysis of water access programmes in Colombia, Furlong (2012) furthers the debate around the political role of infrastructures by pointing to new forms of infrastructure configuration where users and their practices play a central role. In a similar way, this chapter examines the interaction between a new form of infrastructure configuration and the way it seeks to change the consumption practices of users.

A second body of literature, which looks at the connection between infrastructures, consumption practices and public policies aimed at encouraging users to alter their consumption habits in the direction of more so-called 'sustainable' practices, supports taking these ideas forward. The recognition of the growing presence of public policies designed to alter users' consumption habits, primarily for environmental reasons, has led to a significant number of social science studies that explore the connection between policies, instruments for energy efficiency and consumption habits. However, few of these studies have focused on cities of the global South (see Zélem, 2010; Jaglin & Subrémon, 2012). By examining energy management policies, this literature reveals the central importance of a close analysis of practices, revealing the logics of change or resistance. It also points to the need for users – a very heterogeneous group – to be segmented when redesigning political tools intended to encourage changes in consumer habits. For example, policies around what has been termed 'eco-gestures', an approach widely implemented in information campaigns in both northern and southern hemispheres, have been strongly critiqued within the social sciences. According to Jaglin and Subrémon (2012: 1), the relevance assigned to eco-gestures is over-simplistic because it assumes a 'virtuous convergence of small individual actions and isolates practices from their envelope of routines, habits and also constraints'. A substantial number of studies have now demonstrated that energy use is affected by a multiplicity of social and cultural factors (see Subrémon, 2011 for a comprehensive review), and that consumption 'is customary, governed by collective norms and undertaken in a world of things and socio-technical systems that have stabilizing effects on routines and habits' (Shove, 2003: 9). In light of these realities, the eco-gesture approach proves to be of limited value. In contract to such approaches, an emerging body of academic literature within the social sciences points to a more direct connection between infrastructures and social practices without necessarily focusing on a close understanding of individual habit. Shove and Walker (2010: 476), for example, argue in favour of going beyond the introduction of new technologies designed to effect socio-technical transition in order to connect them with 'aspects of practice theory as a means of also conceptualising the dynamics of demand'.

The remaining sections of this chapter offer an in-depth empirical example of the connection between these two bodies of literature. How does an infrastructural reconfiguration with the objective of formalising informal 'archipelagos' engage with consumption practices? To answer this question it is important to focus both on the interactions between the different tools aiming to transform consumers into customers and on the appropriation of these tools by residents in the *favelas* of Rio de Janeiro.

Regularising Electricity Networks in Rio de Janeiro's *Favelas*

Since 1979, the first official electrification programme of *favelas* in Rio de Janeiro has been carried out through the 'social interest electrification programme' (Light, 1980). Introduced by Rio de Janeiro's electricity distribution company, Light, then under federal government control, it officially connected the majority of the city's *favelas* to the public grid. With the installation of individual meters in each dwelling, *favela* dwellers came to be formally recognised as users – just like the city's other inhabitants. It was therefore a substantial political and social achievement, given that until this time formal access to the grid was forbidden to *favela* dwellers on the grounds that land occupancy was illegal. As a result, electricity coverage in Rio de Janeiro appeared to have become universal. For this reason, for at least 15 years, practitioners working in the field have referred to the interventions of Light on the electricity grid as a regularisation of service access. The change in terms, from electrification to regularisation, is not just a question of semantics. What it implies is a change in policy, establishing a policy for managing access to the electricity network. This shift in meaning can be traced back to the 1990s, when the federal government began a wide-ranging restructuring and liberalisation process in the electricity sector.

In Rio de Janeiro, the French firm EDF took a majority holding of Light in 1996.[1] At the time the company faced a complex scenario, as, in a franchise zone with universal network coverage, almost 400,000 dwellings were not customers of the company. Most of them were still connected, arguably illegally, to the grid. Under these circumstances, regularising the grid – or access to the service – meant establishing a plan to develop commercial relations. To this end, between 1999 and 2002 Light implemented a network regularisation programme: the *Programa de Normalização de Áreas Informais* (Programme for the Regularization of Informal Settlements) covered 240 *favelas* and regularised commercial relations for 260,000 customers.

Despite this extensive programme, access to the grid in *favelas* in the 2000s was far from 'regular'. The number of customers registered with the company would vary significantly from one *favela* to another as well as a percentage of the population in a given *favela*. By way of example, the number of registered customers in the Chapéu Mangueira *favela,* situated in Leme, the southern zone of Rio de Janeiro, grew from 430 to 600 after the introduction of the service regularisation process. In the neighbouring *favela,* Babilônia, the same process led to an increase in customers from 414 to 1,017 (Kelman, 2012).[2] The high number of clandestine connections, combined with the company's policy of minimal intervention to improve service quality – justified by the company due to the control of a large section of the *favelas* by armed groups linked with drug trafficking – meant that electricity access in the *favelas* remained precarious in several cases.

The prospect of hosting several major international events (the World Cup and Olympic Games, amongst others) in Rio de Janeiro directly contributed to a change in the way the electricity distribution company operates in the *favelas*. The city's internationalisation prompted the State government to focus its efforts on a new public policy to increase security in the city by maintaining a permanent police presence in certain *favelas*. The UPPs (Pacifying Police Units) have been, since 2008, stationed in *favelas* with the stated objective of re-establishing territorial control. This police presence has established safe working conditions for both government authorities as well as the operators of infrastructural services and other private actors. These actors have exploited this policy as a window of opportunity to establish 'regularised' commercial and non-commercial relations within *favelas*. In order to unpack this window of opportunity, the next section provides a general context for the energy efficiency programme and introduces the actors involved.

The Energy Efficiency Programme in Rio's *Favelas*: 'Regulatory Obligations' and Commercial Activity

A Proactive Role for the State in Promoting Energy Efficiency

Energy represents a crucial resource for the socio-economic development of Brazil. The country's energy management policy has been controlled at the federal level since the 1980s. It continues to be an important policy agenda, driven by several factors. First, rapid industrialisation and a high and sustained rate of economic growth, together with improvements in living standards, have substantially contributed to a profound change in electricity use – despite significant economic disparities between regions and between rural and urban areas. This translates to the rapid growth of the rate of connection to the electricity grid: between 1980 and 2000, the proportion of connected households rose from 68 per cent to 93 per cent (IBGE, 2000). Similarly, total electricity consumption increased between 1975 and 2000 by 250 per cent and average individual consumption by 60 per cent (Geller, 2003: 188). Second, the deterioration of finances in the electricity sector in the 1980s forced the Federal Government to take loans from multilateral development funding agencies with conditions that prompted the sector to look closely at environmental issues and the need to manage electricity demand (De Gouvello & Jannuzzi, 2002).

In 1985 the government introduced the National Electricity Conservation Programme (PROCEL), the country's first electricity demand management programme. PROCEL was based on measures primarily designed to improve production systems and distribution networks (De Gouvello & Jannuzzi, 2002). The programme also targeted end-users, notably through a labelling scheme for electrical appliances with the aim of steering consumer purchasing choices. In addition, the federal government required distribution companies to contribute to the management of energy demand. Federal Law 9,991/2000 mandated electricity distribution companies to invest at least 0.5 per cent of their annual operating revenues in energy-saving programmes (*programas de eficiencia energética* – PEE).

These programmes could target a broad variety of consumers, including local authorities and public services, commercial and service sectors, industry, the residential sector, rural areas and low-income communities. Within this range, low-income communities were considered a priority group, and at least 60 per cent of this investment had to be directed towards users living in these areas.[3] The Brazilian Electricity Regulatory Agency (*Agência Nacional de Energia Elétrica* – ANEEL) monitors the achievement of these goals. As Jannuzzi (2002) notes, the environmental impact of these initiatives is only one of the aspects involved in demand management. The intervention is also designed to achieve increased reliability in the power grid, cut costs associated with electricity production, transmission and distribution, and reduce end-user costs. Against this background, measures designed to promote energy efficiency are combined with the goal of poverty reduction, a primary political objective in Brazil, a country marked by significant socio-economic and spatial disparities.

Actors Implementing the Programme: Private Electricity Providers and NGOs

In compliance with these national level legal obligations, each distribution company conducts energy efficiency programmes in low-income communities. At Light, the *Comunidade Eficiente* programme (Efficient Community) has been in place since 2002, benefiting over 260,000 users in 272 communities (Light, 2011). The programme promotes two types of measures: a technical component consisting of replacing refrigerators and inefficient light bulbs free of charge with appliances that meet national energy efficiency standards; and an 'educational' component, consisting of showing customers how to adopt more temperate energy practices and to avoid energy waste. Combined, these measures seek to promote energy savings in residential electricity consumption.

The implementation of Light's *Comunidade Eficiente* programme was subcontracted to a São Paulo-based company specialising in the implementation of social projects in low-income communities. Although sometimes described as a non-profit by local actors, the organisation is registered as a commercial (for profit) initiative. It is particularly involved in consumer education initiatives, implemented by community agents. These community agents work with users based on an educational working manual provided by Light, teaching basic notions of energy efficiency as well as its economic implications, and explaining the company's pricing framework and methods used for calculating household consumption. In addition, since 2006 Light has been working in partnership with the *Comitê para Democratização da Informática* (Committee for Democratization of Information Technology, or CDI), a non-profit organisation whose mission is to 'strengthen low income communities through the use of information and communication technologies' (*Inclusão* Digital, n.d.; translated from Portuguese). Under this partnership, CDI has introduced the topic of energy efficiency into its digital inclusion programme.

The utility company's aim of promoting energy efficiency within low-income communities also supports the goal of stabilising commercial relations, an objective that the company readily acknowledges: 'Developing educational initiatives focused on the correct and secure use of energy, while seeking to create strategies to reconcile the need for electricity with customers' ability to pay, and to combat irregularities and non-payment' (Light, 2011: 23; translated from Portuguese). Community agents are seen as important intermediaries in the achievement of these multiple objectives. They are not only responsible for the programme's educational component, but also meet face-to-face with customers in the temporary customer service facilities which were installed in some of the *favelas* where the programme was rolled out. In this way, the program's educational measures go hand in hand with the development of customer relations, so that social objectives co-constitute the development of commercial relationships. For example, in order to access the digital inclusion programme delivered by CDI – under its partnership with Light – dwellers were asked to provide an electricity bill and demonstrate that their payments were up-to-date.

By way of summary, the material included in this section illustrates how electricity utility companies respond to ANEEL's efficiency requirements through programmes that target low-income customers whilst developing customer relations and advancing the commercial dimension of service access. The next section unpacks the details of this intersection of objectives, focusing on how energy efficiency measures are used primarily as a tool for stabilising commercial relations and thereby create customers. To achieve this, the normalisation of consumption practices is managed through programmes around consumption control as well as by means of material and commercial socio-technical instruments.

Three Socio-Technical Instruments Used to Change Electricity Consumption Practices in Santa Marta *Favela*

Santa Marta is a *favela* located on a hill in the Botafogo district in the outskirts of one of the city's high-income neighbourhoods, the southern zone (Figure 4.1). It is a medium-sized *favela* with a population of 3,908 people (1,176 households). The majority of inhabitants of Santa Marta live in conditions of poverty: 20 per cent of the area's population has a relatively impoverished socio-economic profile, while 80 per cent belong to what is referred to in official accounts as 'the lower middle class' (IETS, 2010; IPP-Rio, 2013). Socio-economic inequalities relative to Botafogo are significant: whereas in Santa Marta 38.3 per cent of households earn less than one minimum salary per person/month, a legally defined measure of income established by the Brazilian government and corresponding to approximately 160 euros per person/month at the time of fieldwork, only 5.7 per cent of Botafogo's households fall within this salary bracket (IPP-Rio, 2013).

Santa Marta was the first *favela* in Rio de Janeiro to have a Pacifying Police Unit, established in 2008. Since then, it has been a focus of major transformations, including a range of urban development initiatives and the formalisation of all urban services and (previously informal) commercial premises.

500   0   500   1000   1500   2000 m   Image: IBGE RJ25_27454so_V1
                                                             Prepared by Eduardo Bulhões

Figure 4.1 Location of the *favela* Santa Marta, Rio de Janeiro
Source: Eduardo Bulhões

Shaping Consumer and Customer Practices through Socio-Technical Systems

Commencing in 2009, Santa Marta was the first *favela* in which Light chose to introduce an electricity regularisation plan. Before the introduction of the regularisation
plan the company had only 73 customers registered in this area, a significantly low
number when compared to the more than 1,600 customers registered by the end of
the regularisation plan (2010). In other words, clandestine or illegal connections
were the primary mode of electricity access. It is therefore essential to begin by
looking at the methods employed to regularise the electricity grid to better understand how these measures interact with activities to encourage energy savings.

By means of a technical change, the installation of meters and the upgrade of
the electricity network, the company regularised a previously failed commercial
relationship, laying the foundations for a new contractual relationship with users. In
Santa Marta, the technical interventions for regularisation were designed with the
aim of making electricity installations as secure as possible from potential fraud.
This is in contrast to the methods used in *favelas* where there is no Pacifying Police
Unit presence and unsecured overhead cables and mechanical meters are employed.
In Santa Marta, the old distribution network was replaced by a shielded, under-
ground network (33 km of cable). Twenty-one shielded transformers were installed
(Figure 4.2), doubling distribution capacity (Light, 2010a). The old wooden poles
were also replaced by 63 fibre poles, which were more flexible and better suited

Figure 4.2 New power transformers
Source: the author

to the topography of the *favela*. Finally, each user was provided with an individual meter which communicates remotely via a telemetry system with the company's headquarters. Meters are therefore read remotely. Finally, in order to protect the meters from potential fraud, they were not installed in individual dwellings but in 113 armoured cabinets located on the street (Figure 4.3), accessible only by selected company employees (Light, 2010b). These technical measures have led to significant improvements in service quality. Whereas in 2009 there were 29 hours of power failure, in 2011 this figure was reduced to eight hours (Light, 2012). Nonetheless, service quality is still not as good as in the Botafogo district, which in 2011 experienced only 2.56 hours of power failure (ANEEL, 2012).

The characteristics of this supply system point to the role assigned to technical tools in the establishment of the relationship with these new users (customers). As Akrich (1987: 1) explains, technological tools 'define in their configuration a certain division of the physical and social world, allocate roles to certain types of actors – human and nonhuman – while excluding others, [and] allow certain kinds of relationships between these different actors'. In this case, while they are clearly used by the company as a method of preventing commercial losses and indirectly safeguarding the commercial relationship itself, they are also deployed as vehicles for a system of incentives aimed at modifying consumption practices. According

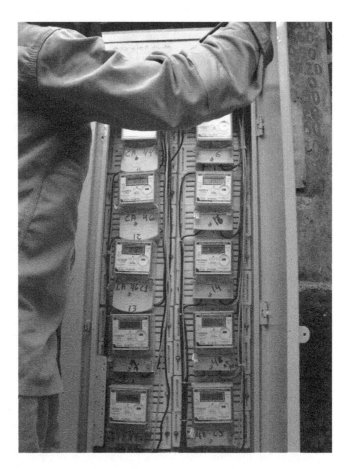

Figure 4.3 Meters in an armoured cabinet
Source: the author

to the company itself, changes in consumption practices are initially encouraged by making people responsible for paying their electricity bills. In this sense, the technical instrument operates prior to the establishment of the new relationship as a tool capable, at least partially, of protecting the new contract against possible 'illicit energy behaviour' (meter tampering or electricity theft via clandestine connections). In this way, it helps to construct the materiality of an incentive to change consumption practices based firstly on the promotion of bill payment, framed in this context as a form of civic behaviour.

Encouraging Consumers to Change Their Consumption Practices: Ability to Pay

Under the programme *Comunidade Eficiente*, refrigerators, light bulbs and other appliances deemed inefficient were replaced at no cost for all households expressing a wish to take part in the programme. In total, around 7,000 fluo-rescent bulbs, 653 refrigerators (42 per cent of households) and 490 electric showers were replaced (Light, 2010a). In addition, teams were deployed to pro-vide information to residents about the methods and advantages of modifying consumption habits.

Encouraging a transformation in practices begins with the injunction to stop 'waste'. This is primarily about promoting the acquisition of more energy-efficient devices which leads to lower energy consumption for the same level of use. It is also about encouraging users to replace their personal, day-to-day energy habits with eco-gestures: 'don't leave the fridge door open if you're not using it', 'turn off the lights when you leave a room', etc. The incitement to change habits is par-ticularly strong in the *favelas*, since consumption practices in those households are often described by the company – and by the media – as 'irresponsible'. Energy consumption practices of *favela* residents are often stigmatised and reported in the media as practices for which society as a whole pays the price, in so far as the cost of commercial losses is partially recovered by the company through an increase in prices for all consumers. However, as acknowledged by the investor relations manager at Light, *favelas* in Rio de Janeiro are responsible for only 40 per cent of commercial losses, while the rest are associated with the 'formal city' (Figo, 2011). In fact, the practice of informally (illegally) connecting to the grid (known in Brazil as *gato*, or 'cat' in Portuguese) is not specific of the *favelas*, neither is it directly related to poverty: the middle and upper classes as well as retailers use this practice to reduce their electricity bill (see Yaccoub, 2010).

As a result, the company's message to its new electricity customers in *favelas* is based around advice to break with the 'superfluous habits' inherited from the period when households did not pay electricity bills. This incitement to energy conservation emphasises self-control and individual responsibility, but also the significant economic advantages associated with it: practising energy con-servation will cut electricity bills. Nonetheless, this message is not free of risk, particularly for low-income users for whom lower energy consumption may be a consequence of economic constraints. The methods employed by the company are thus embedded in economically rational and normative criteria: financial savings and conformity with a new environmental social norm.

As a result, when users experience payment difficulties, they are frequently invited to review their consumption habits, in particular those generally recognised as intensive energy users. For example, the practice of showering is a common topic of advice, especially in the case of large families. Electric showers, whose contribution to total household energy bill can be significant depending on how it is used (hot water, warm water, cold water), are the most common showering appliance in *favelas*. For this reason, intimate and personal questions concerning the use of the electric shower can be asked by the community agents employed by the utility provider. This is illustrated by a dialogue between a company employee and a customer: 'Do you shower every day? In hot water? How many minutes? If you multiply that by the number of the family members, you can see that the shower contributes to a high level of consumption . . . ' (fieldwork notes, October 2011). This dialogue points to the difficulties associated with the attempt to change everyday private practices, highlighting how ability to pay – the economic dimension as a consumption reference for people experiencing payment difficulties – is mobilised towards the definition of the practices which need to be modified. Despite the company's narrative around energy conservation without undermining household comfort, in practice, the boundaries between what is a rational use of energy, energy waste and intensive uses are conflated in the message, in particular when the stability of the commercial relationship seems threatened.

Commercial Tools: A Gradual Change in Modes of Consumption

Progressive payment collection is another way to adjust the consumption practices of customers to their ability to pay. This ad hoc commercial policy, specifically designed for *favelas* with UPP, was introduced in Santa Marta during the first phase of the regularisation project. On the assumption that charging the full cost of energy use from the start of the regularisation process would have negative effects both on collection rates and the project's social acceptability, the company decided to introduce a progressive billing system designed to accustom users in a gradual manner to the idea of consumption-based payment. In practice, for the first six months the amount households paid was set at a consumption ceiling of 79 kWh, even if their consumption was above this ceiling. In subsequent months, the consumption baselines were gradually increased, until reaching a cap of 200 kWh. Only households whose real electricity use was below the established baseline paid for their actual consumption (Table 4.1).

Similar approaches combining commercial and resource management policies have also been used in other contexts, in particular Colombia and Canada (Furlong & Bakker, 2011; Furlong, 2012). Ultimately, according to the utility company, it was the combination of gradual increases in billing amounts with the *Comunidade Eficiente* programme within a technical framework which prevents illegal practices. This was used to gradually nudge households into adopting greater energy-saving practices.

Table 4.1 Progressive method of payment collection in Santa Marta

Period	Consumption paid
July 2009–December 2009	Up to 79 kWh
January–February 2010	Up to 100 kWh
March–April 2010	Up to 120 kWh
May–June 2010	Up to 140 kWh
July–August 2010	Up to 160 kWh
September–October 2010	Up to 180 kWh
November–December 2010	Up to 200 kWh

After this date the billing will be proportional to the actual consumption of each household.

Source: the author, based on fieldwork research (2010)

Household Perceptions: The Lack of Encouragement

The aim of the recommendations provided by Light to users was to modify domestic energy habits by means of two classic energy policy levers: a strong conservation message combining civic dimensions, and the idea of economic rewards resulting from a change in habits. In order to understand how users perceive the regularisation policy and its objective of energy efficiency, a qualitative survey with households in Santa Marta was conducted. The survey was carried out concurrently with the application of the progressive billing policy discussed in the previous section. The results illustrate a poor understanding of the message on energy efficiency, alongside the emergence of a counter-discourse based largely on a critique of the tools employed by the utility company.

'How Much Do I Consume?'

An analysis of the specific tools provided by Light to raise users' awareness of their consumption habits enables a critical understanding of the company's message on energy conservation. Leaving aside simple advice on eco-gestures, discussed in previous sections, this section considers other instruments not specifically identified by the utility company as amongst the range of measures deployed to raise awareness on consumption: the electricity bill and the electricity meter. This analysis steps away from the idea that it is sufficient to provide information on what is deemed to be good consumption practices, focusing instead on the materiality of two specific instruments, the bill and the meter, which operate not only as mediators of the relationship between users and the company but also as devices that inform users about levels and fluctuations in electricity use.

First, the introduction of progressive billing was not accompanied with comprehensive information to help users monitor their real consumption levels. In the case of users whose consumption exceeded the established baseline, only the maximum consumption amount specified in the baseline was itemised on the bill, leaving consumers without any indication of their actual consumption. At the time of survey, 16 out of 25 households consumed more than the baseline (180 kWh). These households did not have access to information on their real level of consumption, or the difference between what was being consumed and what was being billed. Actual consumption levels were displayed (in kWh and monetary value) only on the bills of users whose consumption was below the baseline, and therefore these were the only customers who received information on their real consumption levels.

Second, the meter does not provide a means for tracking consumption. This is because, given the anti-fraud measures outlined earlier, users do not have direct access to their meters. Although electricity meters have been identified as a valuable source of information for both users and utility companies (Furlong, 2007), here they do not perform such a function. Following user complaints, in order to remedy this lack of access to equipment capable of monitoring electricity use, the company distributed electrical devices which, once plugged in, would show the number of kWh used. This device was aimed at smoothing relations between the company and users, by providing greater transparency through information on electricity use. However, these devices failed to achieve the intended objective as they were technically defective. After the discovery of the device's failure, customers complained that the company decided to withdraw them without providing additional information or an alternative mechanism for monitoring consumption.

The ambiguous and partial information given to users led to conflicts between users and the energy provider. Deprived of ways of understanding and tracking their own consumption, users accused the company of billing random amounts (see also Loretti, 2013). The impossibility of checking consumption levels recorded by the meter was condemned by residents in multiple ways. The primary complaint was about the need for people to be able to track their own consumption. In the words of a user, if 'we can't even check our consumption on the meter, it is hard to know how much we consume' (Interview, 2010). To develop a critique of the efficacy of the instruments used by the utility company to encourage energy efficiency, users appropriated the company's message around 'being aware of consumption practices'. Furlong has shown the extent to which 'how meters are managed has direct implications for their capacity to function as instruments to provide incentives for consumers to manage their demand' (Furlong, 2007: 69). In Santa Marta, the ability to be aware of one's own consumption practices (through meter readings) was re-articulated by the public as a necessary step both to achieve the company's objectives and to establish the veracity of the company's billing. Users invoked a denial of access to meters as an expression of the company's lack of trust, accusing it of lacking transparency in order to safeguard its own interests. These arguments show how monitoring consumption is important not only to secure the effectiveness of the message around a change in consumption habits, but also for the construction of a new entente between users and the company.

Metering, billing and electricity payment became a longstanding site of struggle between users and the utility company in Santa Marta. In March 2014, inhabitants of the *favela* took to the streets and blocked the main thoroughfares of Botafogo to protest against the constant increases in electricity bills, with amounts perceived as arbitrary, excessively high and inexplicable relative to the users' domestic appliances (Favela 247, 2014). As identified by researchers working on energy topics from a social science perspective, 'the electricity bill usually functions as an intermediary or unequivocal spokesperson' (Cupples, 2011: 943). In the case of Santa Marta, the information vacuum left by the bill was not filled by a sufficiently comprehensive explanation to 'pacify' the situation. At the time of fieldwork, most of the users interviewed had not understood that the bill only showed the consumption ceiling and not actual consumption, which contributes to misunderstandings regarding the evolution of the billing amount over the following months.

Research on the links between infrastructures and consumption point to the difficulties experienced by users in relating the levels of consumption shown in utility bills and the practices that gave rise to them (Van Vliet et al., 2005). Whilst transparent and accurate information on consumption does not automatically change consumption practices, in the specific context of Santa Marta the construction of users' knowledge on the relationship between level of consumption and practices is even more difficult to experience, because the bill does not provide information on the variation of electricity consumption levels.

Energy Education and Commercial Policies: Complementary or Contradictory?

The commercial policy implemented by the utility company, based on consumption baselines and progressive increases in payment charges, seems to have largely neutralised the economic advantages promised by the message of encouraging people to change consumption practices towards energy conservation. Whereas households expected their bills to reflect their efforts to control consumption, this correspondence could only be verified in the case of households whose consumption was below the established baseline. For all others, the economic incentive was located largely in the future, for the period following the progressive billing system. As a result, although the company presented progressive billing as an additional tool to give users time to adapt to paying bills and to new consumption practices, its impact on the objective of encouraging change in consumption practices towards energy conservation was counter-productive. If the bill is the same no matter what you do, the tendency is to consume more, particularly in the absence of additional information on real consumption. There is thus a contradiction between a message promoting individual responsibility, on the one hand, and a mechanism that, on the other, encourages uniform consumption practices because everyone pays the same. The effect of this configuration, therefore, seems to have been to invalidate the construction of a detailed and realistic understanding of the links between consumption, spending and habits.

As a result, users look for other clues in order to establish whether the level of consumption reported by the company is correct and in this way decide whether it is worth changing habits. In general, users allude to the number of appliances their neighbour has: 'For my part, I'm careful how much I use, I've stopped taking hot showers every day, but it doesn't make any difference, because I still pay as much as my neighbour, who has more appliances than me', says one user (Interview, 2010). Ownership of electrical appliances, rather than consumption habits, is what matters in assessing consumption modes. The sense of injustice aroused by equal payments centres not on the difference in consumption, but on the direct link with appliance ownership, thereby invalidating other determinants that contribute to defining consumption profiles. Relying on the connection between energy savings and financial savings is particularly difficult when it comes to encouraging energy conservation. It would seem all the more problematic in the case of users who have installed a meter for the first time and are therefore starting from scratch in establishing a cognitive connection between consumption and payment. With the progressive billing system, their first experience as customers of the utility company is to see their bills go up continuously.

The extension of the progressive billing system confirms its inefficacy as an incentive to reduce consumption. Whereas, under the timetable initially established by the company the progressive billing policy was set to end in February 2011, it was still going in December 2011. As stated by the Head of Community Relations of the utility company, this decision reflected a perception that 'the expected educational effects have not yet been achieved. A significant number of customers continue to consume more than the baseline. With this level of consumption, we are likely to see an increase in payment defaults' (Interview, 2011).

The empirical material presented illustrates how, in the *favela* Santa Marta, the focus on educating users neglected a concern with habits and custom (Shove, 2003). It also indicates that the messages advocating energy conservation were not backed by real incentivising measures. Despite the utility company's view around the need to approach both consumption practices and commercial dimensions as complementary and connected, the gradual mode of billing reversed the experiential construction of the relationship between consumption and expenditure. In fact, the primary objective of the gradual mode of billing was to increase the acceptability of paying bills, particularly given the low amount charged at the beginning of the regularisation process. Yet, this was done without proven connection between the incentives in place and a reduction in consumption. Such integration of consumption practices and commercial logics risks limiting advocacy around energy conservation to a tool for stabilising commercial relations, therefore potentially aimed at the poorest, and acts to the detriment of a broader understanding of complexity in modes of consumption.

Conclusion

This chapter analysed the shifting relationship between infrastructure reconfiguration and energy conservation policy. It is based on a critical analysis of ANEEL's energy efficiency programme, alongside a more in-depth case study of household electricity use in informal settlements in Rio de Janeiro. It shows how the current wave of regularisation and enhancement of access to electricity services in Brazil is widely accompanied by specific measures that aim to frame and place norms on the consumption practices of low-income users in accordance with service market logics. This suggests that there are particular social and political implications for these forms of infrastructure reconfiguration with respect to the model of inclusion of the low-income population into society. The 2014 protests against growing electricity tariffs discussed above go well beyond the simple demand for comprehensive information around consumption, metering and billing. They raise the issue of the affordability of the service relative to the population's economic profile. These households have experienced years of irregular electricity supply, albeit without paying the full costs associated with such supply. However, now a form of improved access seems rather to encourage them to increase their use of electrical appliances. Yet inhabitants of *favelas* have to deal with the contradictions of energy and economic policies at both national and local levels. On the one hand, both market and public policies encourage users, now framed as consumers, to increase their number of household electrical appliances. While public policies for credit access have been an important tool in fighting poverty and promoting social inclusion in Brazil (Barone & Sader, 2009), the market promotes a consumer society based around commercial credit policies, particularly promoted by shops servicing low-income populations. On the other hand, electricity regularisation encourages this same population to change consumption habits through another quite distinctive conception of energy efficiency oriented around ability to pay. This ongoing and unresolved contradiction appears to place the burden of energy efficiency and conservation squarely on the shoulders of two households types in *favelas*: the poorest, which have global financial restrictions, and the emerging middle class, which has just got out of poverty loops and has now gained widespread access to domestic electrical appliances. In both cases, this contradiction further increases the embedded inequalities within these energy-saving efforts by targeting those who are actually far from being the biggest users of electricity.

Acknowledgements

The author thanks the local actors and the residents who gave up their time to be interviewed. I would also like to thank Jonathan Rutherford and Hélène Subrémon, both from LATTS, for their comments and advice.

Notes

1 In 2006, a 78.4 per cent stake of Light was bought by Rio Minas Energia Participações S.A.
2 It should be emphasised, however, that these numbers diverge from those of the 2010 census. According to the 2010 census, Chapéu-Mangueira has 401 dwellings and Babilônia 777, numbers which are lower than those declared by the electricity company.
3 At the time of our survey, the Energy Efficiency Programme was targeted at 'low-income communities', so all residential users living in these areas were potential beneficiaries. However, following the passing of law 12.212/2010, the target group was modified. After this law came into force in 2011, only users meeting the criteria for access to the subsidised social tariff would be entitled to the programme.

References

Akrich, M., 1987. 'Comment Décrire les Objets Techniques?'. *Techniques & Culture* 9: 49–64.
ANEEL, 2012. *Informações Técnicas: Indicadores de Continuidade*. [Online]. Available from: http://www.aneel.gov.br/area.cfm?id_area=80 [Accessed: 30 August 2012].
Bakker, K., 2003. 'Archipelagos and Networks: Urbanization and Water Privatization in the South'. *The Geographical Journal* 169: 328–41.
Barone, F.M. and E. Sader, 2009. 'Access to Credit to Fight Poverty in Brazil'. *International Journal of Case Method Research & Application* 21(1): 19–25.
Coutard, O., 2008. 'Placing Splintering Urbanism: Introduction'. *Geoforum* 39(6): 1815–20.

Cupples, J., 2011. 'Shifting Networks of Power in Nicaragua: Relational Materialisms in the Consumption of Privatized Electricity'. *Annals of the Association of American Geographers* 101(4): 939–48.

De Gouvello, C. and G. Jannuzzi, 2002. 'Maîtrise de la demande d'électricité et secteurs électriques publics monopolistiques: Comparaison France – Brésil'. *Annales des Mines* 31–8.

Favela 247, 2014. Santa Marta protesta contra contas da Light. *Jornal Brasil 247.* [Online]. 26 March 2014. Available from: http://www.brasil247.com/pt/247/favela247/134579/Santa-Marta-protesta-contra-contas-da-Light.htm. [Accessed: 27 March 2014].

Figo, A., 2011. Pacificação de *favelas* no Rio ajuda a Light ampliar área de cobertura e elevar resultado. *Infomoney.* [Online]. 24 February. Available from: http://www.infomoney.com.br/mercados/noticia/2024880/pacificacao-*favelas*-rio-ajuda-light-ampliar-area-cobertura-elevar-resultado. [Accessed: 3 March 2014].

Furlong, K., 2007. *Municipal Water Supply Governance in Ontario: Neoliberalization, Utility Restructuring, and Infrastructure Management.* A Thesis Submitted in Partial Fulfilment of the Requirements of University of British Columbia for the Degree of Doctor of Philosophy. Vancouver: University of British Columbia.

Furlong, K., 2012. Mediating the Gap Between Custom, Cost-recovery, and "Networks": An Example from Colombia. In *From Networked to Post-networked Urbanism: New Infrastructure Configurations and Urban Transitions.* Autun (France), Tuesday 17 to Friday 20 July 2012. p. 25.

Furlong, K. and K. Bakker, 2011. 'Governance and Sustainability at a Municipal Scale: The Challenge of Water Conservation'. *Canadian Public Policy* 37(2): 219–37.

Geller, H.S., 2003. *Revolução Energética – Políticas Para um Futuro Sustentável.* Rio de Janeiro: Relume Dumarà.

Graham, S. and S. Marvin, 2001. *Splintering Urbanism: Networked Infrastructures, Technological Mobilities and the Urban Condition.* New York: Routledge.

IBGE, 2000. *Censo demográfico de 2000.* Rio de Janeiro: Instituto Brasileiro de Geografia e Estatística.

IETS, 2010. *Pesquisa nas favelas com Unidades de Policia Pacificadora da Cidade do Rio de Janeiro. Resultado Consolidado.* Rio de Janeiro: IETS. [Online]. Available from: http://www.iets.inf.br/article.php3?id_article=1769 [Accessed: 15 June 2011].

Inclusão Digital, (n.d.) *Comunidades.* [Online]. Available from http://www.ong-inclusao-digital.org.br/comunidades.htm [Accessed: 7 April 2013].

IPP-Rio, 2013. *Panorama dos Territórios UPP SANTA MARTA – UPP Social.* [Online]. Rio de Janeiro: IPP-Rio. Available from: http://www.riomaissocial.org/territorios/santa-marta/. [Accessed: 25 April 2013].

Jaglin, S., 2005. *Services d'eau en Afrique subsaharienne. La fragmentation urbaine en question.* Paris: CNRS éditions.

Jaglin, S. and H. Subrémon, 2012. La transition énergétique à l'épreuve des logiques d'usages: le cas des petites classes moyennes au Cap. In *Premières journées internationales de sociologie de l'énergie.* Toulouse, p. 13.

Jannuzzi, G., 2002. Aumentando a Eficiência nos Usos Finais de Energia no Brasil. In *Sustentabilidade na Geração e o Uso da Energia no Brasil: os próximos 20 anos.* São Paulo, Monday 18 to Wednesday 20 February 2002. [Online]. Available from: http://www.bibliotecadigital.unicamp.br/document/?code=1035&opt=1 [Accessed: 6 May 2015].

Kelman, J., 2012. The Long and Winding Road. Rio Slums & the Electric Utility. In: *Harvard Brazil Studies Program.* Harvard, Thursday 19 April. p. 38. Available from: http://www.kelman.com.br/pdf/22JKinHarvard.pdf [Accessed: 6 May 2015].

Kooy, M. and K. Bakker, 2008. 'Technologies of Government: Constituting Subjectivities, Spaces, and Infrastructures in Colonial and Contemporary Jakarta'. *International Journal of Urban and Regional Research* 32(2): 375–91.

Light, 1980. *Eletrificação de favelas no Municipio do Rio de Janeiro.* 3 Convenção Brasileira de Assistentes Sociais. Rio de Janeiro: Light.

Light, 2010a. *Relatório de Sustentabilidade 2009.* Rio de Janeiro: Light.

Light, 2010b. *Relatório anual Responsablidade Socioambiental 2009.* Rio de Janeiro: Light.

Light, 2011. *Manual de trabalho educativo. Programa comunidade eficiente, Light.* Rio de Janeiro: Light.

Light, 2012. *Relatório de Sustentabilidade 2011.* Rio de Janeiro: Light.

Lopes De Souza, M., 2005. *O desafio metropolitano: um estudo sobre a problemática sócio- espacial nas metrópoles brasileiras.* Rio de Janeiro: Bertrand Brasil.

Loretti, P., 2013. "Atirei o pau no gato": da regularização da energia elétrica aos mecanismos de controle e repressão no contexto de "pacificação" da favela Santa Marta. In XVI Congresso Brasileiro de Sociologia, Salvador de Bahia, p. 31.

Lorrain, D. (ed.), 2011. *Métropoles XXL en pays émergents.* Paris: Presses de Sciences Po.

McFarlane, C., 2008. 'Governing the Contaminated City: Infrastructure and Sanitation in Colonial and Post-Colonial Bombay'. *International Journal of Urban and Regional Research* 32(2): 415–35.

McFarlane, C. and J. Rutherford, 2008. 'Political Infrastructures: Governing and Experiencing the Fabric of the City'. *International Journal of Urban and Regional Research* 32(2): 363–74.

Shove, E., 2003. *Comfort, Cleanliness and Convenience: The Social Organization of Normality.* Oxford: Berg.

Shove, E. and G. Walker, 2010. 'Governing Transitions in the Sustainability of Everyday Life'. *Research Policy* 39(4): 471–6.

Subrémon, H., 2011. *Anthropologie des usages de l'énergie dans l'habitat. Un état des lieux.* Paris: Editions Recherche du PUCA.

USAID, 2009. *Transforming Electricity Consumers into Customers: Case Study of a Slum Electrification and Loss Reduction Project in São Paulo, Brazil.* Washington, DC: Agency for International Development.

Van Vliet, B., H. Chappells and E. Shove, 2005. *Infrastructures of Consumption: Environmental Innovation in the Utility Industries.* London: Earthscan Publications.

Yaccoub, H., 2010. *Atirei o pau no "gato". Uma analise sobre consumo e furto de energia elétrica (dos "novos consumidores") em um bairro popular de Sao Gonçalo – RJ.* Programa de Pós-graduação em Antropologia, Niteroi: Universidade Federal Fluminense. MSc Thesis.

Zélem, M.-C., 2010. *Politique de maîtrise de la demande d'énergie et résistances au changement: Une approche socio-anthropologique.* Paris: L'Harmattan, Logiques sociales.

5 Delhi

Questioning Urban Planning in the Electrification of Irregular Settlements

Laure Criqui

Introduction: Unplanned yet Electrified

It is generally considered that traditional urban planning in the global South is deficient and, as such, makes the extension of basic services problematic or complicated (Davis, 2006). At first sight, unplanned settlements present irregularities in terms of spatial organisation, social regulation and regulatory enforcement, often seen as challenges for the extension of utilities (Criqui, 2015). Based on empirical data from Delhi, India, this chapter questions this assumption. By pointing to the ways in which the electricity distribution companies that operate in the city extend electricity networks into unplanned settlements, this research analyses the mechanisms used to compensate for inadequate urban planning, providing insights into their impact on the urban fabric, development and governance. Caught in the tension between intervening in 'illegal' settlements and expanding service coverage, the process of electrification in Delhi follows three main trends. First, using a pragmatic logic, utility companies muddle through unplanned settlements using devices and practices particularly adapted to irregular conditions. Second, they try to articulate the physical extension of infrastructures within a chaotic urban fabric and politicised governance. Third, they adopt differentiated strategies that reveal the highly political yet unacknowledged impact of utilities' interventions in developing irregular settlements.

The chapter starts from the premise that infrastructures are socio-technical, material and political objects that play a critical role in shaping the configuration of the urban (McFarlane & Rutherford, 2008). It contends that their extension is a process worth analysing through those very social, material and political prisms (Rutherford & Coutard, 2014). The materiality of infrastructure extension, and its impact on the consolidation of irregular settlements, not only shapes the city but also sheds light on the geographical dimensions of the electricity grid and its functioning (Bridge et al., 2013). Seeing infrastructure deployment as both a material and a political process highlights the interdependence between service extension, the urban fabric and politics.

Methodologically, the chapter uses semi-structured interviews with utility executives, engineers and technicians to uncover the mechanisms shaping the urban interventions of electricity distribution companies in Delhi. This material affords insights into their daily work practices and *modus operandi*. A review of the regulatory literature, project reports and corporate communications material helps to establish the background of the electricity companies' actions. In addition, interviews with public officials and members of civil society situate the electrification process within larger urban development dynamics in the city. Finally, the methodological approach is complemented with site visits, which enabled an on-the-ground assessment of the scope and constraints of the extension of the electricity grid through an irregular urban fabric.

Urban Growth and Planning in Delhi

As the National Capital Territory of India, Delhi occupies a unique position in the country: it is the largest urban agglomeration of India, with a status similar to that of any Indian state and with its own specific governance structure. Established in 1992, the Delhi government is an independent executive authority in charge of, among others, urban development, water and sanitation, and electricity. In addition, five local boards and municipalities manage civic services. The Delhi Development Authority, a central government body created in 1957, was established to promote and secure the development of the capital. Having adopted a traditional master planning approach – based on restrictive land-use and zoning regulations to control urban growth – it has been overwhelmed by the rapid growth that has taken place over the second half of the 20th century (Datta & Jha, 1983; Nath, 1995; Ahmad et al., 2013; Bhan, 2013). In 1941 Delhi had a population of 900,000. This figure doubled over the course of one decade and reached 22 million by 2011.

Delhi's master plan distinguishes three categories of settlements: planned areas which host the middle and upper classes and represent 25 per cent of the city (New Delhi, British colonial zones and sections of South Delhi); the special area of the Old City, characterised by congestion, mixed-land use and areas with heritage value, which accounts for around 20 per cent; and over 50 per cent of the city considered to be unplanned areas. Within the latter, different sub-categories account for a variation in rights – and levels of access – to public services, programmes and policies. Half of unplanned Delhi consists of permanent and consolidated 'unauthorised colonies' as well as urban and rural villages now engulfed by urban sprawl. These settlements have been developed on land bought from agricultural owners or are directly occupied, sub-divided and built-up in breach of planning regulations and building by-laws (Banerjee, 2002; Risbud, 2002). Since 1993, the Delhi government has been trying to tackle the issue of regularising and bringing development works and service provision to these unauthorised colonies and villages, with multiple delays and difficulties (Zimmer, 2012; Lemanski & Tawa Lama-Rewal, 2013). The other half of the unplanned city consists of small pockets of slums and squatter settlements – where housing and environmental precariousness prevail (Baviskar, 2003; Dupont, 2008; Bhan, 2009) – as well as remote, precarious and under-serviced resettlement colonies for evicted slum dwellers (Ali, 1995).

Accounting for approximately 12 million people, unplanned settlements in Delhi are key areas of intervention for both the government and utility companies, be it for electoral purposes, the development of new market opportunities or the achievement of social development goals. It is in this unplanned city where the population is not completely connected to services, thus where utility companies have room to extend their network and improve access. Yet, it is also where they have to adapt their practices to work in irregular conditions.

Electricity Reform in Delhi

Until 2002, electricity was managed by the Delhi Vidyut Board, part of the Delhi government. Insufficient coverage and frequent blackouts plagued electricity distribution, and theft tolerance and poor management kept the sector in a vicious cycle of low-performance (Sagar, 2004). Facing growing discontent, the Delhi government reformed the sector following a now conventional model of unbundling services and opening them up to private capital (Wamukonya, 2003). The reform was underpinned by the creation of the independent Delhi Electricity Regulatory Commission (DERC). For the purpose of electricity distribution Delhi's territory is divided into five zones: New Delhi and the Cantonment remain under the special jurisdiction of their own local municipality and board, whilst the remaining 95 per cent of the city is serviced by three private utility distribution companies, commonly referred to as the Delhi DISCOMs. The DISCOMs are joint ventures, where 51 per cent of the shares are owned by private Indian conglomerates – controlling strategy and management – and 49 per cent by the Delhi government (Srivastava & Kathuria, 2014). Each DISCOM has a monopoly over its concession zone: BSES Rajdhani Power Limited (BRPL) and BSES Yamuna Power Limited (BYPL), both owned by the Reliance Group, operate respectively in the South and West and in the central and East zones; Tata Power Delhi Distribution Limited operates in the North (TPDDL) (Figure 5.1).

Studies show that utility performance and electricity distribution quality in Delhi have improved since 2002 (Srivastava & Kathuria, 2014). The sector has also been working on increasing access, and according to the 2011 census, electricity connection stands at 92.9 per cent. The DISCOMs have adopted what they refer to as a 'techno-commercial strategy': first, upgrading formerly dilapidated infrastructures to bring better service quality to consumers who are already connected, which secures revenue. Then, following the stabilisation of technical and commercial losses (Figure 5.2) from 2007 onwards, the DISCOMs started to extend the electricity network to new consumers, specifically to irregular settlements (Figure 5.3).

In evaluating the effect of Delhi's electricity reform, both academic and political debates have focused on regulatory and politico-economic aspects: tariff setting and the efficiency of the new companies, the level of autonomy of the regulator and the interests and power balances in the new governance system (Dubash & Rajan, 2001; Stamminger, 2002; Agarwal et al., 2003; Mahalingam et al., 2006; Thakur et al., 2006; Dubash & Rao, 2008; Bhide et al., 2010; Kasturi, 2013; Srivastava & Kathuria, 2014). Little interest has been paid to the way the DISCOMs have practically and concretely transformed the distribution system to extend the grid: the

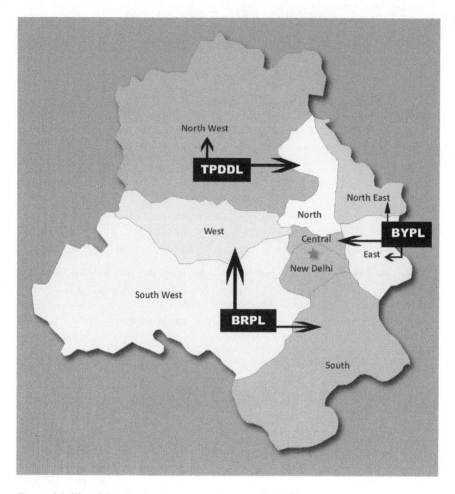

Figure 5.1 Electricity distribution concession zones in Delhi
Source: the author

operational translation of the reform is largely unknown. The technical and commercial challenges of upgrading an existing network, setting up infrastructures in
unconventional spaces and incorporating a fragile client-base are under-studied,
despite the extent to which effective tools and implementation mechanisms for servicing irregular settlements constitute critical factors for such reforms to succeed.

The Extension of Networks Despite Irregularity

The gap between the share of irregular settlements (around 55 per cent) and
the rate of connection to electricity (over 90 per cent) in Delhi suggests that the
DISCOMs provide services to unplanned settlements; it means that an absence

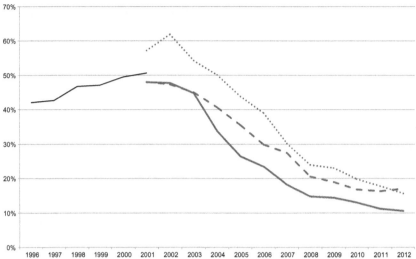

Figure 5.2 Aggregate technical and commercial losses
Source: the author, based on BRPL, BYPL, TPDDL, DERC annual reports

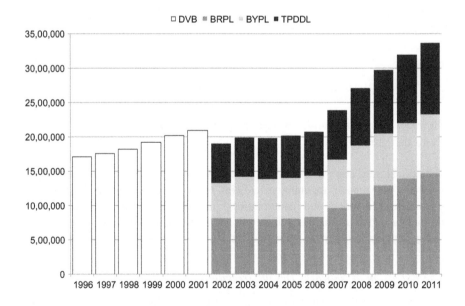

Figure 5.3 Total number of electricity consumers in Delhi
Source: the author, based on BRPL, BYPL, TPDDL, DERC annual reports

of formal planning is not *per se* an obstacle to electrification. In this context, it is important to understand how utilities tackle the legal, social and spatial challenges of irregular settlements in their attempts to extend their network (Criqui, 2015). Such a viewpoint provides an opportunity to unpack effective practices of urban development outside the realm of official planning, focusing on their material impact and political meaning.

Socio-Technical Innovations for the Connection of Irregular Settlements

The rationale for electrifying irregular settlements is two-fold. The DISCOMs enter irregular areas prone to high rates of theft to curtail distribution losses but this is also an opportunity to expand their market base. To do so, they resort to a series of solutions that help them overcome the consequences of infringement of the urban plan and rules. As explained by the CEO of a DISCOM,

> *if I do not bring services, people will steal. That is a financial loss and a technical risk for my network. It will burden the honest paying consumer, and that also creates social discontent. So it is a better strategy to connect them and make them pay.*

(Interview, 2012)

Locational Knowledge and Institutional Bricolage

Being unplanned, irregular settlements do not appear on official plans and no clear regulatory framework applies. For utility companies, one of the key challenges is to get them 'on the map', in order to know where to intervene, manage the demand to be serviced and stabilise the intervention framework. When urban planning is irrelevant, utilities resort to a series of 'informal' institutional rules (Leitmann & Baharoğlu, 1998). This 'institutional bricolage' relies both on traditional tools and new arrangements (Cleaver, 2001; Merrey & Cook, 2012), and gets its efficiency and legitimacy from its embeddedness in the informality of urbanisation itself (McFarlane, 2012). Following that logic, the Delhi DISCOMs have produced their own set of rules and procedures which act as institutions.

First, they have simplified the prerequisites to apply for a connection: an ID and residency proof replace the requirement of a property title. The proof of residency can be a letter from either the local elected representative or the residents' association; despite the 'illegal' character of the settlement, its actual existence is legitimated by these official sanctions. Second, on receiving applications from the clients-to-be, the DISCOMs need to locate the premises to be electrified, a difficult task when there is neither address nor map to identify them. To do so, utilities' employees walk through settlements painting on each door a 'K-number' that systematically combines lane and house numbers (Figure 5.4) – a technique borrowed from the rural *Khasra* land registry system. Linking together the house's K-number, the client's contract number and the registration details of a newly

Figure 5.4 Electricity company's K-number and outside meter encased in sealed box
Source: the author

installed individual electricity meter, the DISCOMs can locate the electrified premises and follow up with service delivery. The DISCOMs have thus developed their entire customer base and network connections on the basis of their own "informal" identification and address system.

Such an exercise of recognition through – and alongside – the creation of addresses is a prerequisite for the DISCOMs to overcome an institutional lack of knowledge about irregular settlements. In doing so, they create a new service provision framework that relies on pre-existing legitimate institutions transformed into pragmatic tools – e.g. elected representatives and traditional land registry systems. In locating electricity demand, the DISCOMs organise the settlements, an ordering of informal space and through this a way of institutionalising it. The resulting institutional 'bricolage' (Cleaver, 2001) has political effects of its own: the K-number and the electricity contract formalise relations between a usually marginalised population and the DISCOMs. Such recognition of informality, not sanctioned but yet tacitly accepted by public authorities, silently locates Delhi's informal settlements on the public map and agenda. Though privatised, the DISCOMs are formal organisations in which the state maintains a high stake. Furthermore, in settlements where public presence is weak, utility companies are considered state representatives by the people (Akrich, 1987). In Delhi, for example, residents feel more confident they will not be evicted when they can show an electricity bill. It provides a form of tenure security and a proof of address that enables access to other public or private services; hence it becomes a first step in the administrative and thus political recognition of the settlement.

The actions of the DISCOMs in Delhi point to the emergence of rules and institutions for the purpose of governing utilities which are outside of the realm of official plans and regulations. As observed in Istanbul (Leitmann & Baharoğlu, 1998), sidestepping official planning rules, bricolage systems and informal mechanisms underpin the intervention made by utilities, actors that are not concerned with urban illegality.

Commercial and Social Arrangements

The development of a wide and reliable range of customer services operates as a mechanism to guarantee bill recovery by increasing clients' willingness to pay (Caseley, 2006). Through customer service, utility companies not only improve the accessibility and accountability of service provision (World Bank, 2004), but can also spread civic and moral discourses aimed at turning their clients into responsible consumers who do not default on payment. With the ultimate aim of reducing commercial losses, the DISCOMs have invested in satisfying the Delhiites' expectations and in this way convince them it is worth paying for electricity services. Since 2002, the DISCOMs have put in place comprehensive customer care systems, offering 24/7 phone lines as well as a diversity of payment schemes, such as neighbourhood office, Internet, text message and doorstep collection services. The automation of meter reading, invoice distribution and payment collection suppresses human intervention, reducing the possibility of corruption or extortion and increasing clients' trusts in the company's management systems.

An example of the implications of minimising human intervention on the grid is provided by the discontinuation of the Single Point Delivery mechanism for unmetered slums, a system previously used by the Delhi Vidyut Board. The Single Point Delivery was a connection given to a local headman, who was responsible for the technical and financial management of distribution within the settlement. The system, abandoned in 2007 when the DISCOMs started electrifying irregular settlements, is highly criticised by DISCOMs' managers who see it as inefficient and open to corruption, where service access was based on personal relations. In order to keep technical and financial control over the network, the DISCOMs favour direct relations with their clients. Thanks to individual metering and contracting, each client is now personally responsible for their connection and interacts directly with the companies, reducing opportunities for corruption or political manipulation.

Lastly, the DISCOMs have implemented social programmes aimed at improving customers' willingness and ability to pay for electricity services. TPDDL, for example, established a corporate social responsibility strategy based on professional training, drug-addiction recovery camps, health care and women literacy classes, among others. Access to these services is conditional on the presentation of paid invoices. Such corporate social responsibility approaches operate as a two-sided strategy: whilst following the Tata Group's philanthropic credo of 'giving back to the people', it transforms the poor into 'honest and responsible' consumers via targeted incentives.

Through a new type of relationship with the population, the DISCOMs have attempted to instil client respect towards the infrastructure and service offered. Based on the regulator's annual reports and statistics,[1] the DISCOMs have achieved a level of 100 per cent efficiency in bill collection from 2006 onwards. This means that any remaining distribution losses are due to technical rather than commercial challenges.

Anti-theft Technical Devices

Through a search for technical efficiency and a fight against electricity theft, utilities in Delhi have transformed conventional electricity networks (Dasgupta & Atmanand, 2013; Banerjee & Pargal, 2014). Within the space of a decade, new devices, materials and techniques have changed Delhi's electricity landscape. From a technical perspective, the main new device is the High Voltage Distribution System (HVDS), made of proximity transformers hung on distribution poles, directly converting high- to low-tension in the immediate vicinity of the nodes of consumption (Figure 5.5). HVDS reduces technical losses in two ways. First, transformers are sealed; and since the in-feed is a high-tension current, 'hooking' to the grid in an informal manner becomes more dangerous and difficult. Second, the length of low-tension lines is reduced, further limiting distribution losses. Through this device, the DISCOMs have increased network safety, improved service to tail-end customers and reduced electricity theft. In highly dense urban areas where the installation of HVDS is not possible, the DISCOMs use low-tension aerial bunched cables and

Figure 5.5 High Voltage Distribution System transformer
Source: the author

constantly look for new ways of armouring cables to prevent tapping. Tampering with the system is also prevented at the point of metering, through the installation of sealed boxes for new generation meters on the outside walls of the dwellings, a strategy that allows the meter to remain visible (Figure 5.4). Automated meter reading is also part of this technical search for reinforcing equipment.

According to the DISCOMs' managers and engineers, the initial resistance to metering and billing – exercised by a population used to getting free (illegal) electricity – was overcome thanks to an increase in service quality. The positive effects of formal electricity access and improved service quality are ratified by local residents, a change that has lowered a perceived sense of discrimination in service access for low-income populations. Differentiation in service provision can at times be considered a form of spatial, social or even political discrimination and provoke resistance (Jaglin, 2008). But in Delhi, urban planning practices already generate sociospatial segregation (Bhan, 2013). In a culture of exclusion, the differentiation of service provision per category of settlement follows rather than creates an existing splintered urbanism (Zérah, 2008). Such embedded forms of splintering are counterbalanced by improved access to the grid.

The DISCOMs have managed to electrify unplanned settlements through a combination of technical, commercial and institutional bricolage. In their view, any legal, social and spatial irregularities can be overcome as they consider that their work is precisely to find solutions leading to an expansion of the utilities' activities and networks. In the words of an engineer, urban irregularity does not present problems; rather, the focus is on the generation of solutions: 'Technically we do not face urban problems, we can find solutions for everything, we can give the service without difficulty, that's our job!' (Interview, 2012). This apparently neutral and efficiency-oriented techno-commercial extension of the grid has nevertheless silent effects on the political recognition, integration and formalisation of irregular settlements. Service extension in irregular settlements is particularly meaningful given their lack of inclusion in the official spheres of urban planning. Through installing poles, arranging contracts with customers and creating addresses, utilities participate in the process of urban development (Akrich, 1987), a process that is both spatial and political (McFarlane & Rutherford, 2008; Rutherford & Coutard, 2014).

Coordination Constraints on the Deployment of Electrical Networks

Governance deficits, typical of Delhi's political economy (Pinto, 2000; Siddiqui et al., 2004), influence the DISCOMs' ability to preserve their infrastructure from urban development dynamics. As innovative as they may be, utilities still have to muddle through an urban fabric that is not always appropriate for infrastructure deployment (Connors, 2005; Criqui, 2015). In the absence of an urban plan, public works are haphazard and collide or damage each other. On the ground, coordination inefficiencies and physical externalities from different sectorial logics reflect the internal tensions of Delhi's governance. The lack of planning – the exercise through which urban interventions can be made coherent – appears to be a challenge difficult for the DISCOMs to overcome. They thus have to incorporate these externalities and inefficiencies into network management. Nevertheless, these conflicting processes have long-lasting deteriorating effects on the urban fabric, particularly affecting irregular settlements where public authorities do not orchestrate the urban development process. As highlighted by the CEO of a DISCOM, 'the problem is not the electrification of unauthorised colonies, that's easy! The real challenge and difficulties come afterwards' (Interview, 2013).

Encroachments, Public Space and Safety Distances

The lack of urban planning alongside little enforcement of local by-laws creates spatial difficulties in securing space for infrastructure. The first challenge is related to a limited definition of public space and to conflicts over priorities around its use: streets, a particularly scarce resource in fast-growing cities, are spaces of competition between buildings, construction and infrastructures (Marvin & Slater, 1997). Since the 1990s, Delhi's unplanned but consolidated unauthorised colonies and villages have been becoming increasingly dense. This trend has intensified since the mid-2000s with the government's campaign for regularisation and thus coincides with the entry of the DISCOMs into irregular settlements. As a result, it has become increasingly critical – and difficult – to preserve space for infrastructures.

To deploy their network, the DISCOMs require space for new transformers and poles. Furthermore, the increase in energy demand results in the need for additional sub-stations. However, the density of unplanned urbanisation, with residential uses encroaching on public areas and road widths, forces the DISCOMs to look for new locations for placing infrastructure. The first option is to put transformers and sub-stations on the berm of roads, often the only easily accessible land (Figure 5.6). Another option is to install stations on peripheral roads rather than inside the settlements, surrounding them with a high-tension network and then pulling the distribution wires within. These solutions are, however, not optimal, since they are technically unsafe and unsecured from and for traffic. Moreover, the institutional implementation is full of complexities: depending on the size, land category, zoning and other variables, Delhi's roads are the responsibility of over seven different public agencies. The DISCOMs are required to obtain authorisation from these agencies in order to set up their infrastructures, resulting in complex, long and costly procedures.

Once infrastructures are installed, the built-up environment constantly encroaches upon them, threatening their proper functioning (Figure 5.7). The DISCOMs have to respect horizontal and vertical safety distances, yet have neither the competency nor the time to enforce building by-laws requiring the demolition of third-party constructions breaching the minimum distances required around electricity equipment. Whilst public authorities are supposedly responsible for enforcing these norms, they do not do so in practice. The DISCOMs are forced to resort to other coping tactics, such as locating electrical equipment further away from constructions through, for example, on-arm elevated or pole-mounted transformers (Figure 5.9) or armoured and rigid low-tension aerial bunched cables. This increases the distance between high-tension systems and buildings or between two poles, reduces the possibility of illegal connections and ensures that electricity infrastructures are not threatened by traffic.

Figure 5.6 Sub-station on the berm of roads
Source: the author

Figure 5.7 Encroachment by a balcony
Source: the author

The Street Lighting Mess

Street lighting, seen as an 'anomaly' by the regulator, provides a typical example of Delhi's complex governance system, where the logics of elected politicians, public administration and corporate utilities collide. Street lighting in Delhi is normally the responsibility of road-owning agencies, up to seven depending on settlement category, land status, road width, etc. They establish the lighting lay-out, decide on any extensions, and establish a contractual relationship with the DISCOMs to build and operate the network. Technically, the DISCOMs operate exclusively as executing agencies with no responsibilities for feasibility evalua-tion. As external contracted bodies, they do not have the opportunity to integrate this street lighting work with the planning, implementation or maintenance sched-ules of their own local distribution strategies. Electrification and street lighting are thus two independent processes with significant potential for overlapping or colliding. The complexity of the arrangements between the DISCOMs and the various government agencies responsible for roads and streets prevents the proper follow-up and maintenance of street lights. High-level managers from the DISCOMs recount a long-lasting dispute between local authorities and DISCOMs around the allocation of responsibilities and payments, contributing to a deficient and erratic operation of the system (Figure 5.8).

Furthermore, in practice, street lighting is ordered by elected representatives who, independently from public intervention schemes, use discretionary funds for local development projects for this purpose. In this way, politicians have the power to favour particular urban settlements, neighbourhood blocks or streets as per their vote-seeking priorities (de Wit, 2009; Keefer & Khemani, 2009). Access to street lighting occurs in a piecemeal manner, depending on politicians' influ-ence and support. In a context of night insecurity (Viswanath & Mehrotra, 2007), and with the background of corruption scams associated with street lighting for the 2012 Commonwealth Games (Baviskar, 2014), political and administrative confusion around street lighting have become acute.

Public Works in Haste

Delhi's complex governance also creates significant difficulties for the coordina-tion of public works at street-level. The absence of coordination between over 20 public, parastatal and private agencies in charge of delivering public works and services negatively impact on the existing equipment of the grid. Over-ground works create conflicts around the use of space. Since sub-stations are located on street berms (Figure 5.6) and the DISCOMs are not informed of operation and maintenance works, the risk of flooding of electrical equipment is high. More-over, as part of settlement regularisation programmes and in the context of the 2013 local elections, elected representatives extensively supported road paving, a process that affected electric infrastructures. Roads are paved by adding succes-sive layers of different materials, raising street levels up to 1 metre high. Besides burying houses, road paving buries electricity poles thus lowering transformers relative to ground level and reducing safety distances (Figure 5.9).

Figure 5.8 Street lighting in use during daytime
Source: the author

Figure 5.9 On-arm transformer and pole buried by road paving
Source: the author

Underground urban processes also affect electricity infrastructure, such as deficient drainage, water and sanitation systems generating frequent street flooding, and threatening electricity equipment installed on berms. Underground works are erratic and destabilise the physical environment in which infrastructures are entrenched. In Delhi, these normally follow the principle of 'deposit-and-dig', through which the agency conducting works is supposed to compensate for potential damages by providing the road-owning agency with the necessary budget to repave the street once works have been completed. DISCOMs' executives argue that financial under-estimation, delays in interventions, low quality of repairs, lengthy procedures and neglect prevent any proper and sustainable functioning of this principle. Trench digging is a blind exercise performed by each sectorial agency in its own way and following its own priorities. The rush for public works for political purposes, mainly a function of electoral cycles, damages the electricity network in visible ways.

Differentiated Approaches for Dealing with Unplanned Urbanisation

At first sight, the electrification of irregular settlements could be considered a neutral and uniform process of reducing technical and commercial losses. For all three DISCOMs involved, similar socio-technical innovations and coping strategies appear to compensate for deficient planning. Further analysis, however, reveals significant differences in each DISCOM's approach, pointing to a multiplicity of logics at play (Criqui & Zérah, 2015). Since 2007, the DISCOMs have tackled the socially and politically sensitive tasks of engaging with vulnerable populations, being accepted as legitimate service providers, getting by-laws enforced and contracts implemented in order to fulfil their mission. Each DISCOM has adopted different approaches to achieve this, revealing different ways of defining and relating to the public interest through the electricity grid. These differences become apparent in the diverging public positions of each company and the way in which the specific socio-technical characteristics of the concession zones are dealt with (Criqui & Zérah, 2015).

Diverging Positioning of the Companies

Through an analysis of DISCOMs' corporate communication strategies it is possible to identify diverging understandings of what servicing irregular settlements means. The increased media attention that follows utility privatisation processes puts new service providers under the spotlight, increasing the amount of information available to the public (Caseley, 2006). An evaluation of the Delhi DISCOMs' monthly newsletters provides evidence of the roll-out of a form of civic governmentality (Watson, 2013) aimed at taking responsibility and self-disciplining, in this way shaping the mind-sets and behaviours of the newly established consumers. The alignment of utilities' priorities and people's expectations creates a cognitive path-dependency (Cleaver, 2002; Hall & Lobina, 2007), and imposes a

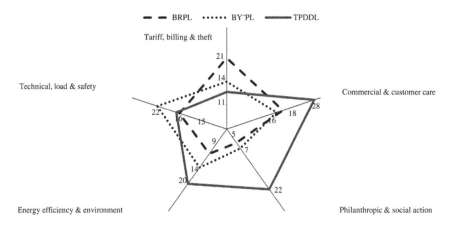

Figure 5.10 Thematic items of DISCOMs' newsletters
Source: the author, based on BRPL, BYPL, TPDDL's newsletters[2]

set of dominant lenses for interpretation: economic and financial priorities, technical and efficiency concerns, and social responsibility and accountability (Figure 5.10) (Criqui & Zérah, 2015).

BRPL's newsletters and statements during public Delhi Electricity Regulatory Commission hearings and research interviews reveal a specific position around tariffs and financial questions. In the case of all three DISCOMs, 80 per cent of the company's costs come from power purchase; tariffs are set by the regulator based on a consideration of such costs. BRPL particularly complains about the absence of financial margins for investments, is reluctant to take any action for which costs are not officially approved in advance and criticises government subsidies to the poorest consumers. The company also denounces electricity theft and payment defaults, whilst systematically blaming low tariffs for putting a strain on its financial equilibrium and investment capacity. BRPL positions itself as the victim of a system that constrains its capacity to generate the profits required for higher levels of investment in electricity infrastructures. This position is expressed by a company executive, who highlights the financial restrictions that the company experiences in its interactions with customers:

> *I am private company, I cannot change the world; and I am only a retailer: as per the tariff system, whatever expenditure is done, and whatever profit I make, my profits are regulated . . . if I make more money, it reduces the tariff for the next year. If I make an expenditure, it goes on the tariff. Whatever money I collect, it goes to the consumers. No profit is retained by any DISCOM. DISCOM is only like a retail shop, we do not have control over prices. And we depend on other states for generating power. So if their rates go up, I cannot do anything.*
>
> (Interview, 2013)

Besides, the window of opportunity that privatisation usually offers for increasing technical and energy efficiency (Dyner & Larsen, 2001) does not seem to have been seized by Delhi's DISCOMs. The lack of involvement from both public authorities and civil society does not favour energy efficiency or demand management measures (Zérah & Kohler, 2014). Beyond anti-theft solutions that bring immediate results, there is little research and development of innovative devices or new technological solutions. During research interviews with all three DIS-COMs, BYPL revealed a unique concern with issues of technical efficiency for intervention in unplanned settlements. The CEO appeared committed to R&D promotion and building partnerships with technology institutes in Delhi. In line with this commitment, BYPL has developed intelligent monitoring systems for sub-stations and solar water pumps, and in 2013 initiated research around pre-assembled transformers that would make installation and retrofitting easier.

TPDDL appears particularly committed to improving social relations. The company established a dedicated corporate social responsibility cell and a 'special consumer group' to engage with customers in irregular settlements. The company's corporate social responsibility cell works with local NGOs to implement social activities and hires local residents to serve as intermediaries with the population. Its staff members are particularly proud of this form of social involvement and Delhiites positively acknowledge the efforts made. The Delhi government and civil society movements value the character of TPDDL as a socially responsible DISCOM – a feature that comes across both in the media and through research interviews – by drawing attention to the type of commercial initiatives undertaken by the company. By contrast, they criticise the lack of attention to social issues provided by BRPL and BYPL.[3]

The diverging orientations used by all three companies reveal differences in the rationalities used to conceptualise electrification and energy access. How can the differences in the DISCOMs' agendas be explained?

Sociospatial Characteristics of Concession Zones

The co-existence of different strategies stemming from a unique privatisation reform reveals the extent to which utilities' interventions are spatially constituted (Bridge et al., 2013; Criqui & Zérah, 2015). Delhi's urban landscape is characterised by its sociospatial fragmentation, a phenomenon that is reflected in the sociospatial challenges utilities face.

BRPL's concession zone is one of contrasts. Upper and middle-classes live in the planned neighbourhoods of South Delhi and peripheral estates. Already connected to the grid, this population is not so much concerned with the extension of networks as with the quality of the service (Baviskar, 2003). Civil society activists living in these areas mobilise towards increased transparency and better client services, and complain about theft and illegal connections. Yet unauthorised colonies are widespread within the South-western fringe of the concession zone and are considered to be "high [electricity]-theft prone areas" (Zimmer, 2012). Therefore the BRPL's concession zone is characterised by social tensions which could explain the company's discourse around the need for responsible (electricity paying) consumers. Its constant condemnation of electricity theft and illegal connections can be interpreted as an attempt to satisfy the 'honest paying consumers' who do not default on payment.

The spatial characteristics of the BYPL's area make it a very specific environment to intervene in: it is highly urbanised, almost fully unplanned, dense, congested and engulfed by urban sprawl. The critical issue for BYPL is not extending a network that already covers the entire area, but increasing its capacity to provide services by using the existing infrastructure network. Under these conditions, the company's challenges are manifold. These include a lack of available space for infrastructure expansion – which foregoes the installation of new equipment – and an increased energy demand resulting from both urban densification and the continuous growth in households' electrical appliances. In addition, illegal connections from the neighbouring Eastern State, where BYPL has no license, threaten the overall stability of the network. These spatial constraints force BYPL to enhance its network capacity, an intervention for which technical improvements and energy efficiency are helpful.

The North concession, under the responsibility of TPDDL, concentrates the most economically deprived districts of Delhi (Baud et al., 2008) as well as the large majority of resettlement colonies for slum dwellers. Planned by the Development Authority, these colonies enjoy security of tenure and follow a regular urban layout. But they also host what is defined in India as the economically weaker population, the poorest of the poor, accounting for approximately 10 per cent of TPDDL's consumer base. As a result of this social configuration, a primary area of concern for TPDDL is to secure this population's ability and willingness to pay. This is done through corporate social responsibility actions as well as other initiatives carried out by its commercial facilities.

Last but not least, the fact that the DISCOMs focus on financial, technical or social dimensions of electrification is far from politically neutral or exogenously determined. Though it does not relate directly to the irregular or informal character of unplanned settlements, there are also internal factors and deliberate choices that shape the DISCOMs' strategies. The Reliance and Tata groups to which they belong have specific identities, histories and strategies that reflect on their organisational cultures and management and the companies' societal commitment is also decisive in shaping Delhi's electrical landscape (Criqui & Zérah, 2015).

Conclusion: Electrification and Urbanisation

Delhi's traditional planning system excludes the majority of the city's urbanised area. This formal exclusion blurs the range of possibilities of intervention by a multiplicity of urban actors, particularly utility companies tasked with the delivery of public services. Caught in the tension between respecting the urban master plan or following informal forms of urbanisation, the DISCOMs have opted for extending service coverage, muddling through the city's complex and chaotic urban fabric (Connors, 2005; Criqui, 2015). In foregrounding the institutional and socio-technical work of the DISCOMs towards the electrification of Delhi's irregular urbanisation, this chapter points to their social and political characteristics. The extension of the formal electricity grid, a process deeply embedded in the urban fabric, illustrates a way of overcoming planning deficiencies in the city. Electricity companies take on a significant relevance as urban actors, with electrification as a form of urban development.

Rigid and land-use focused master planning – as sanctioned by the Delhi Development Authority – is inapplicable on the ground (Jain, 2008) and drives service providers away from the procedures formally established by planning authorities (Kumar, 2000). In practice, such limited flexibility and lack of recognition of the informality of the city encourages the development of ad hoc logics for service delivery. Since the city's formal planning tools do not provide objectives or intervention rules appropriate to the reality of the city, the DISCOMs resort to their own procedures, institutional instruments and socio-technical tools in a pragmatic way, irrespectively of the planning framework. Whilst the sociospatial characteristics of electricity concession zones determine the scope and extent of the challenges experienced by utility companies, the presence of diverging strategies demonstrate that each utility also has room to manoeuvre. These differentiated economic and managerial strategies reveal specific visions and interpretations of public interest and priorities. The choices utility companies make in approaching unplanned settlements shape the way people access basic services, and more generally the way Delhi is progressively consolidated, built and developed. These electricity utility companies are not to be seen as neutral actors (Marvin & Guy, 1997; Criqui & Zérah, 2015); they purposefully elaborate their own visions and techniques for urban development, based on their specific operational constraints, interests and strategies (Criqui, 2015). What is, therefore, the impact of such autonomous 'unplanned' developments on the fabric and functioning of the city?

In Delhi, the extension of the grid carried out by the DISCOMs, as a form of infrastructure deployment in unplanned settlements, is considered not only possible but more importantly legitimate. Formal electrification has provided unplanned settlements with a sign of recognition positively affecting those populations excluded from the city's master plan. The DISCOMs have diversified both the techniques and devices enabling service provision. Whilst service differentiation is usually criticised for its contribution to splintering urbanism (Jaglin, 2008), in Delhi, already entrenched in an urban culture of differentiation (Zérah, 2008), grid extension follows rather than guides fragmentation. Undeniably, there are different patterns between formal and unplanned settlements: the multiplication and overlapping of urban development projects in unauthorised colonies, largely due to government regularisation announcements and the lobby of elected politicians, generates disorder and inconveniences both for inhabitants and utility providers. The governance deficits and the lack of coordination for interventions in unplanned settlements make network extension more difficult and less efficient than in planned colonies. The city's recent electrification process illustrates that coordination deficiencies in planning in Delhi may be more decisive for urban development than differentiation itself.

This chapter makes a call for considering electrification beyond a simple search for efficiency, but in the context of an urban development process with crucial social and political dimensions. Largely as a result of privatisation, which transferred the responsibilities of electrification from the state to private companies and thus freed the process from public administration and planning constraints, electricity access in Delhi has played an indirect but important role in the government's agenda to regularise unauthorised colonies. In contrast, water and sanitation access – under government responsibility – has remained largely unequal (Roy, 2013), inadequate and insufficient (Ranjan Panda & Agarwala, 2013). It is argued that the government board in charge cannot override planning rules as easily as privatised companies. In the case of electricity, the government's commitment to sector reform has supported the DISCOMs' intervention. As a result, public development works take place in unauthorised colonies partly through the action of non-state actors. A supportive policy environment has made the formal electrification of Delhi's unplanned settlements possible and tacitly authorised, benefiting the population in spite of a restrictive master plan (Kundu, 2004). As the first official actors to enter irregular settlements, electricity distribution companies have a decisive role and infrastructure extension shapes both the material fabric and the sociospatial integration of unplanned settlements. This intervention in irregular settlements shows the under-studied political and spatial importance of electrification as a means of urban development. Through this process, utility companies can be seen as key actors working under a neutral technical and commercial cover to advance urban development goals neglected by urban policy. The consolidation of formal electrification in Delhi, beyond a technical intervention, should be seen as a material consolidation of unplanned urbanisation, a governance framework that shapes the relations between the population and official authorities, and a sociopolitical intervention resulting in transforming the city's conditions of inequality and enhancing the rights to the city through service provision.

Notes

1 Delhi Electricity Regulatory Commission's annual reports 2003–2012: http://www.derc. gov.in/Publications/Publications.html
2 Based on 21 BRPL's, 17 BYPL's and 32 TPDDL's one-page newsletters from January 2011 to December 2013, taking into account the headlines of the 3–5 main articles.
3 http://www.thehindubusinessline.com/opinion/the-power-sectors-dark-spot/article 5222014.ece; http://timesofindia.indiatimes.com/city/delhi/Arvind-Kejriwal-threatens-to-throw-out-BSES-discoms/articleshow/30010907.cms

References

Agarwal, M., I. Alexander and B. Tenenbaum, 2003. *The Delhi Electricity DISCOM Privatization: Some Observations and Recommendations for Future Privatization in India and Elsewhere*. Washington, DC: World Bank.
Ahmad, S., O. Balaban, C.N.H. Doll and M. Dreyfus, 2013. 'Delhi Revisited'. *Cities* 31: 641–53.

Akrich, M., 1987. 'Comment Décrire les Objets Techniques?'. *Techniques & Culture* 9: 49–64.

Ali, S., 1995. *Environment and Resettlement Colonies of Delhi.* New Delhi: Har-Anand Publications.

Banerjee, B., 2002. Security of Tenure in Indian Cities. In: A. Durand-Lasserve and L. Royston (eds.), *Holding their Ground: Secure Land Tenure for the Urban Poor in Developing Countries.* London: Earthscan/James & James, pp. 37–57.

Banerjee, S.G. and S. Pargal, 2014. *More Power to India: The Challenge of Electricity Distribution.* Washington, DC: World Bank.

Baud, I., N. Sridharan and K. Pfeffer, 2008. 'Mapping Urban Poverty for Local Governance in An Indian Mega-city: The Case of Delhi'. *Urban Studies* 45(7): 1385–412.

Baviskar, A., 2003. 'Between Violence and Desire: Space, Power, and Identity in the Making of Metropolitan Delhi'. *International Social Science Journal* 55(175): 89–98.

Baviskar, A., 2014. Dreaming Big: Spectacular Events and the 'World-class' City: The Commonwealth Games in Delhi. In: J. Grix (ed.), *Leveraging Legacies for Sports Mega-Events.* New York: Palgrave Macmillan, pp. 130–41.

Bhan, G. 2009. '"This is No Longer the City I Once Knew". Evictions, the Urban Poor and the Right to the City in Millennial Delhi'. *Environment & Urbanization* 21(1): 127–42.

Bhan, G., 2013. 'Planned Illegalities. Housing and the "Failure" of Planning in Delhi: 1947–2010'. *Economic & Political Weekly* 48(24): 58–70.

Bhide, S., P. Malik and S.K.N. Nair, 2010. *Private Sector Participation in the Indian Power Sector and Climate Change.* Paris: Agence Française de Développement.

Bridge, G., S. Bouzarovski, M. Bradshaw and N. Eyre, 2013. 'Geographies of Energy Transition: Space, Place and the Low-carbon Economy'. *Energy Policy* 53: 331–40.

Caseley, J., 2006. 'Multiple Accountability Relationships and Improved Service Delivery Performance in Hyderabad City, Southern India'. *International Review of Administrative Sciences* 72(4): 531–46.

Cleaver, F., 2001. 'Institutional Bricolage, Conflict and Cooperation in Usangu, Tanzania'. *IDS Bulletin* 32(4): 26–35.

Cleaver, F., 2002. 'Paradoxes of Participation: Questioning Participatory Approaches to Development'. *Journal of International Development* 11: 225–40.

Connors, G., 2005. 'When Utilities Muddle Through: Pro-poor Governance in Bangalore's Public Water Sector'. *Environment & Urbanization* 17(1): 201–18.

Criqui, L., 2015. 'Infrastructure Urbanism: Roadmaps for Servicing Unplanned Urbanisation in Emerging Cities'. *Habitat International* 47: 93–102.

Criqui, L. and M.-H. Zérah, 2015. 'Lost in Transition? Comparing Strategies of Electricity Companies in Delhi'. *Energy Policy* 78: 179–88.

Dasgupta, M. and N.A. Atmanand, 2013. 'Efficiency in Indian Electricity Distribution through Technological Innovation'. *International Journal of Indian Culture & Business Management* 6(4): 477–90.

Datta, A. and G. Jha, 1983. 'Delhi: Two Decades of Plan Implementation'. *Habitat International* 7(1–2): 37–45.

Davis, M., 2006. *Planet of Slums.* New York: Verso.

De Wit, J., 2009. 'Municipal Councillors in New Delhi: Agents of Integration or Exclusion?'. Paper Presented N-AERUS Conference Rotterdam, 1–3 October. pp. 1–25.

Dubash, N.K. and S.C. Rajan, 2001. *The Politics of Power Sector Reform in India.* Washington, DC: World Resources Institute.

Dubash, N.K. and D.N. Rao, 2008. 'Regulatory Practice and Politics: Lessons from Independent Regulation in Indian Electricity'. *Utilities Policy* 16(4): 321–31.

Dupont, V., 2008. 'Slum Demolitions in Delhi Since the 1990s. An Appraisal'. *Economic & Political Weekly* 43(28): 12–18.

Dyner, I. and E.R. Larsen, 2001. 'From Planning to Strategy in the Electricity Industry'. *Energy Policy* 29(13): 1145–54.

Hall, D. and E. Lobina, 2007. 'Profitability and the Poor: Corporate Strategies, Innovation and Sustainability'. *Geoforum* 38(5): 772–85.

Jaglin, S., 2008. 'Differentiating Networked Services in Cape Town: Echoes of Splintering Urbanism?'. *Geoforum* 39(6): 1897–906.

Jain, A.K., 2008. 'Urban Land Policy and Management Reforms'. *Institute of Town Planners, India Journal* 5(3): 60–8.

Kasturi, K., 2013. 'Pricing Electricity in Delhi'. *Economic & Political Weekly* 48(1): 20–3.

Keefer, P. and S. Khemani, 2009. *When Do Legislators Pass on 'Pork'? The Determinants of Legislator Utilization of a Constituency Development Fund in India.* Washington, DC: World Bank.

Kumar, A., 2000. 'Some Problems in the Co-ordination of Planning: Managing Interdependencies in the Planning of Delhi, India'. *Space and Polity* 4(2): 167–85.

Kundu, A., 2004. 'Provision of Tenurial Security for the Urban Poor in Delhi: Recent Trends and Future Perspectives'. *Habitat International* 28(2): 259–74.

Leitmann, J. and D. Baharoğlu, 1998. 'Informal Rules! Using Institutional Economics to Understand Service Provision in Turkey's Spontaneous Settlements'. *Journal of Development Studies* 34(5): 98–122.

Lemanski, C. and S. Tawa Lama-Rewal, 2013. 'The "Missing Middle": Class and Urban Governance in Delhi's Unauthorised Colonies'. *Transactions of the Institute of British Geographers* 38(1): 91–105.

Mahalingam, S., B. Jairaj and S. Naryanan, 2006. *Electricity Sector Governance in India: An Analysis of Institutions and Practice.* Washington, DC: World Resources Institute.

Marvin, S. and S. Guy, 1997. 'Infrastructure Provision, Development Processes and the Co-production of Environmental Value'. *Urban Studies* 34(12): 2023–36.

Marvin, S. and S. Slater, 1997. 'Urban Infrastructure: The Contemporary Conflict Between Roads and Utilities'. *Progress in Planning* 48(2): 247–318.

McFarlane, C., 2012 'Rethinking Informality: Politics, Crisis, and the City'. *Planning Theory & Practice* 13(1): 89–108.

McFarlane, C. and J. Rutherford, 2008. 'Political Infrastructures: Governing and Experiencing the Fabric of the City'. *International Journal of Urban and Regional Research* 32(2): 363–74.

Merrey, D. and S. Cook, 2012. 'Fostering Institutional Creativity at Multiple Levels: Towards Facilitated Institutional Bricolage'. *Water Alternatives* 5(1): 1–19.

Nath, V., 1995. 'Planning for Delhi and National Capital Region: Review of Plan Formulation and Implementation'. *Economic & Political Weekly* 30(35): 2191–202.

Pinto, M.R., 2000. Delhi's Governance. In: *Metropolitan City Governance in India*. Delhi: Sage Publications, Ltd. pp. 129–61.

Ranjan Panda, G. and T. Agarwala, 2013. 'Public Provisioning in Water and Sanitation Study of urban slums in Delhi'. *Economic & Political Weekly* 48(5): 25–42.

Risbud, N., 2002. Policies for Tenure Security in Delhi. In: A. Durand-Lasserve and L. Royston (eds), *Holding Their Ground: Secure Land Tenure for the Urban Poor in Developing Countries*. London: Earthscan/James & James, pp. 59–74.

Roy, D., 2013. 'Negotiating Marginalities: Right to Water in Delhi'. *Urban Water Journal* 10(2): 97–104.

Rutherford, J. and O. Coutard, 2014. 'Urban Energy Transitions: Places, Processes and Politics of Socio-technical Change'. *Urban Studies* 51(7): 1353–77.

Sagar, J. (2004) Reforms in Electricity. In: 3iNetwork (ed.), *India Infrastructure Report*. New Delhi: Oxford University Press, pp. 162–73.

Siddiqui, K., N. Ranjan and S. Kapuria, 2004. Delhi. In: K. Siddiqui (ed.), *Megacity Governance in South Asia*. Dhaka: The University Press Limited. pp. 89–274.

Srivastava, G. and V. Kathuria, 2014. 'Utility Reforms in Developing Countries: Learning from the Experiences of Delhi'. *Utilities Policy* 29: 1–16.

Stamminger, M., 2002. *Privatisation of Electricity in Delhi*. New Delhi: Center for Civil Society.

Thakur, T., S.G. Deshmukh and S.C. Kaushik, 2006. 'Efficiency Evaluation of the State Owned Electric Utilities in India'. *Energy Policy* 34(17): 2788–804.

Viswanath, K. and S.T. Mehrotra, 2007. ' "Shall We Go Out?" Women's Safety in Public Spaces in Delhi'. *Economic & Political Weekly* 42(17): 1542–8.

Wamukonya, N., 2003. 'Power Sector Reform in Developing Countries: Mismatched Agendas'. *Energy Policy* 31(12): 1273–89.

Watson, V., 2013. 'Planning and the "Stubborn Realities" of Global South-East Cities: Some Emerging Ideas'. *Planning Theory* 12(1): 81–100.

World Bank, 2004. *Making Services Work for Poor People*. Washington, DC: World Bank.

Zérah, M.-H., 2008. 'Splintering Urbanism in Mumbai: Contrasting Trends in a Multilayered Society'. *Geoforum* 39(6): 1922–32.

Zérah, M.-H. and G. Kohler, 2014. 'Le déploiement des énergies propres à Delhi aux prises avec la défiance de la société urbaine'. *Flux* 3(93): 31–42.

Zimmer, A., 2012. 'Enumerating the Semi-visible: The Politics of Regularising Delhi's Unauthorised Colonies'. *Economic & Political Weekly* 67(30): 89–97.

6   Maputo

Fluid Flows of Power and Electricity – Prepayment as Mediator of State-Society Relationships

Idalina Baptista

Introduction

Prepayment systems are increasingly popular in Africa for the delivery of urban utility services, especially for electricity and water. The case for prepayment of utility services has been made particularly relevant in the context of the Millennium Development Goals' focus on improving access to quality services by the poor (Briceño-Garmendia et al., 2004), climate change debates involving the energy poor (Casillas & Kammen, 2010; Prins et al., 2010), and concomitant affordability discussions in infrastructure and tariff reform (Pachauri et al., 2004; Sagar, 2005; Fankhauser & Tepic, 2007). Much of the support for prepayment comes from an eclectic group of economists and development and energy specialists who see the technology as a potential solution to the problems of the energy poor (Estache et al., 2002; Tewari & Shah, 2003; Casarin & Nicollier, 2008). They tend to highlight the advantages of prepayment for both utility service providers and users in contexts of weak governments, scant infrastructure planning, unclear land tenure and persistent poverty. They suggest that prepayment facilitates the expansion of access to utilities in low-income areas, empowers customers (especially the poor), generates revenue to service providers and improves their relationship with users. Prepayment is offered as an appropriate means to improve the efficiency of utility billing of low-income customers without a steady income, where there may be low levels of literacy, high rates of utility pilfering, dispersed or irregular settlement and an inadequate street address system and/or postal service (Tewari & Shah, 2003). Prepayment is said to help households manage their disposable income and to stay out of debt by consuming only what they can afford, when they can afford it (Estache et al., 2002; Casarin & Nicollier, 2008).

In spite of the accolades, scholarship in urban studies has remained particularly critical of prepayment. While much of the existing literature reports on the different experiences within the UK and South Africa (e.g. Marvin & Guy, 1997; McDonald & Ruiters, 2005; McDonald, 2009), the anti-prepayment stance in both contexts seemingly shares a common argument. In particular, critics see the adoption of the technology as representing the reversal – if not the demise – of the public service ethic and its 'modern infrastructural ideal' due to a turn to neoliberalism (Graham & Marvin, 2001). However desirable this 'modern infrastructural ideal' and its public service ethic may remain today, it is questionable the extent to which it has been a reality in many cities of the South (Kooy & Bakker, 2008). In Sub-Saharan Africa alone, where service provision has historically involved a complex array of social actors (Blundo & Le Meur, 2009), it is unclear whether the anti-prepayment stance provides a powerful critique of, or potentially simplifies and obscures, the diverse geographies and techno-politics of urban infrastructure provision and consumption.

In fact, the technologies and devices through which utility services are provided, such as prepayment meters, are central to how people experience the city. They are not neutral technical objects. Technologies and devices are invested with a specific politics about who the utility users are, what kinds of lives they lead and their political subjectivity (Akrich, 1992). It is for this reason that prepayment technology can have the disciplining effect identified by some scholars (e.g. Schnitzler, 2008) which may enforce strict economic relations between users and service providers in the first instance (Loftus, 2006; Heusden, 2009). Prepayment meters can also constitute a political terrain, as argued by Schnitzler (2013), for the negotiation of broader ethical, cultural, environmental and even spatial relations implicated in the organisation of social life (Barry, 2001; Bijker et al., 2012; Marres, 2012).

This chapter examines how prepayment operates as a mediator of state-society relationships in the context of the broader politics of urban infrastructure provision in the cities of the South, particularly in Sub-Saharan Africa. Using the case study of prepaid electricity in Maputo, Mozambique, this chapter examines the ways in which prepayment reconfigures everyday electricity consumption practices and how it elicits an understanding about state-society relationships. The chapter draws on archival research and fieldwork conducted in Maputo in 2013 and 2014. Fieldwork consisted of participant observation and semi-structured interviews with 30 households in nine neighbourhoods of Maputo, 25 informants at six vending shops and with 14 informants related to the electricity sector. Findings from the Maputo case study support a complex and fluid view of the work Prepayment meters do in reconfiguring electricity consumption and of how users envision that consumption in relation to their urban experience. Moreover, findings suggest that electricity users' everyday engagement with prepayment elicits a critical positionality vis-à-vis the state and their condition as citizens. As a result, the chapter reflects on how the current trend towards adoption of prepayment across the African continent may result in a particular politics of urban infrastructure service provision.

The chapter is organised in four further sections. The following section examines existing literature on prepayment, with a particular attention to critiques within urban studies that address state-society relationships in regard to utility service provision. An overview of electricity in Mozambique in general and introduction to the case study in Maputo is then provided, which is followed by an examination of the practices of prepaid electricity in different neighbourhoods of Maputo. The chapter concludes with a discussion of how the case of Maputo expands questions about the everyday governance of urban life in Sub-Saharan Africa and the challenges these raise to theories of urbanisation.

Prepayment of Utility Services: Critiques from within Urban Studies

Cities have been privileged places for the strategic negotiation of claims to citizenship between urban dwellers and the state (Holston & Appadurai, 1996). Historically, urban infrastructures have been deeply implicated in this process of enrolling (or excluding) people in the project of nation-building (Scott, 1998; Graham & Marvin, 2001). In colonial contexts, the provision of urban infrastructures served to reinforce relations of rule between colonisers and colonised, to establish variegated forms of subjectivity and citizenship, which translated to a spatially fragmented landscape of access to basic services (Kooy & Bakker, 2008). Change has been slow in postcolonial contexts, especially in Sub-Saharan Africa. Recent trends towards adoption of prepayment as the retail method for utility service delivery are picking up in many countries and contributing to a change in the geography of urban energy and water politics across the African continent. A cursory online survey conducted in 2014 by the author showed that 43 African countries had adopted prepayment as the retail model for electricity and/or water provision, even if only as part of pilot projects. Prospects of a profitable market for Prepayment technology are enticing investors on the back of wider narratives of smart metering and smart cities.[1]

Most of the critiques of prepayment available in urban studies literature focusing on Africa have been developed from analyses of prepayment in South African townships. Hence, drawing on these critiques to examine the case of prepayment in Maputo later in this chapter requires that we take stock of the specificities of the South African context and the limits of such comparative conceptual work. On the one hand, South Africa has been leading the development of prepayment technology and its implementation since the 1980s (Schnitzler, 2013). The expansion of prepayment throughout the African continent, including Mozambique, is intimately connected with the growth and development of the South African prepayment industry. On the other hand, we can hardly dissociate existing critiques of the South African case from the specificities of post-Apartheid struggles over national reconstruction. Protests against prepayment in townships have recurred in the last two decades (Schnitzler, 2008, 2013). They tend to figure prominently in debates about citizenship rights in relation to access to basic public services (McDonald & Ruiters, 2005). While similar struggles are less visible in other countries where the politics of urban infrastructure has been constituted historically in a different way, the contention between utility service providers and users unfolds along similar lines. In Mozambique, for instance, public criticism to the poor quality of service provided by the state-owned electricity company EDM (*Electricidade de Moçambique, E.P.*) often makes headlines, but there is hardly any criticism targeting prepayment itself (see below). With these possibilities and limitations in mind, this chapter examines the case of Maputo, Mozambique, in tandem with existing critiques of prepayment in South Africa as a useful starting point for expanding our understanding of how prepayment is shaping the geography of urban service provision elsewhere in Sub-Saharan Africa.

Despite the diverse disciplinary origins and theoretical inclinations, scholars have largely converged on a critique of prepayment around three key aspects: (a) prepayment as a proxy for neoliberalism; (b) prepayment as a disciplining technique; and (c) prepayment as de-politicising state-society relationships. These are examined in turn below.

Prepayment as a Proxy for Neoliberalism

The deployment of prepayment has been heavily criticised in South Africa where it is associated with uneven, unequal and unfair processes of neoliberal privatisation and commodification of essential public goods, like water and electricity (e.g. McDonald & Ruiters, 2005; McDonald, 2009). Prepayment is said to facilitate the implementation of harsh full cost-recovery policies (McInnes, 2005), often imposing disadvantageous prices on the poor (and black), when compared to the prices and quality of the services which the rich (and white) access (Flynn & Chirwa, 2005; McDonald & Ruiters, 2005). Some authors go as far as to argue that prepayment reflects a selective form of violence and punishment by the South African state over some of its 'unruly' (poor) citizens who have a culture of non-payment (McInnes, 2005; Naidoo, 2007; Ruiters, 2007, 2009, 2011). This criticism has been essential in scrutinising the (mostly) negative effects of policies targeting privatisation and commodification of collective resources, especially when said policies place paramount importance on the economic viability of service provision at the expenses of social equity. Some scholarship has been more cautious in interpreting the ambiguous effects of the use of prepayment meters in South African townships (Plancq-Tournadre, 2004; Jaglin, 2008). In particular, Jaglin (2008) has noted that the anti-prepayment stance reflects a wider attitude against technical differentiation in utility provision. She argues this attitude may miss the progressive potential of technical differentiation in responding to the diversity of urban livelihoods in developing cities.

In fact, an over-commitment to neoliberalism as a conceptual and analytical lens may be a less profitable avenue for theorising prepayment of urban services in Sub-Saharan cities than for cities of the North at the core of urban theory (Robinson, 2006; Ferguson, 2009; Parnell & Robinson, 2012; Baptista, 2013). In foregrounding the structural relationships between service providers and consumers, the focus on the turn to neoliberalism risks missing the extent to which, in cities of the South, utility supply has historically involved public, private and community providers (Coutard, 2008; Blundo & Le Meur, 2009). In the African context, as Simone (2004) would argue, people tend to be *the* infrastructure through which individual households access basic urban services and secure other necessary resources essential to their provisional and uncertain livelihoods. In fact, 'the state' remains an elusive and contested reality in many African countries, where notions of the 'public/private' divide or 'citizenship' are entangled with forms of allegiance and subject-making which are different from those presupposed by the model of Western democracies (Simone, 2000; Blundo & Le Meur, 2009; Robinson, 2011). Instead of taking the relationship between utility service providers and consumers for granted, it seems more profitable to examine how that relationship is constituted and contested in everyday practices of utility consumption.

Prepayment as a Disciplining Technique

In fact, it is through the everyday use of a utility that its specific technology comes to mediate the relationship between utility service providers and users. This is why some urban scholars have criticised prepayment as a technique for remote disciplining and social control, especially of the poor (Marvin & Guy, 1997; Ruiters, 2007, 2009, 2011). Drawing insights from South African townships, critics deconstruct claims of consumer empowerment as a mode of governmentality designed to inculcate calculation as the essential attribute and ethos of the modern and rational citizen (Harvey, 2005; Schnitzler, 2008). Putting an explicit emphasis on the relationship between individuals and technology, the prepayment meter emerges in these critiques as a pedagogical device that restricts, rather than encourages, the possibilities of self-determination and action. In other words, prepayment's potential for empowerment is said to be an excuse for imposing an economic calculability and metrological scrutiny unwanted by users (Schnitzler, 2008).

Scrutiny of prepayment in technological terms provides useful insights into how the introduction of a new metering technology can reframe the relationship between utility service providers and users. Yet, as Akrich (1992) would argue, assuming that technical objects such as prepayment meters pre-determine that relationship or are, by contrast, merely neutral handmaidens to a relationship of domination defined by providers alone, misses the ways in which technical objects are active participants in that relationship. On the one hand, by anticipating who its users are, what kinds of lives they lead and their political subjectivity (Guy & Marvin, 1995), prepayment meters contribute to define a framework of action for both utility service providers and users (Akrich, 1992). On the other hand, because users may come to subvert those anticipated roles or put technical objects to unanticipated uses as they consume the utility, prepayment meters become a focal point for investigating the shifting nature of the relationship between utility users and providers. For instance, Schnitzler (2013) reports how utility users in Soweto and Johannesburg pulled out prepayment meters from the buildings where they had been installed and dropped them at the door of local government offices in an act of protest and defiance. In her examination of these contentious situations, Schnitzler (2013) recognised that prepayment is inevitably enrolled in a specific historical, political and ethical assemblage that may be constituted differently outside of South Africa. As a result, when examining the case of prepaid electricity in Maputo, Mozambique, it may be profitable to remain open to scrutinising the various ways in which prepayment meters can be actively involved in the organisation of social life and how this engagement may be constituted differently.

Prepayment as De-politicising State-Society Relationships

The contested nature of the relationship between utility service providers and users has been brought up already in the context of the two previous critiques of prepayment. Where the service provider is intimately associated with the state – as in the cases of both South Africa and Mozambique – the provider-user relationship is often conceptualised as a proxy for state-society relationships as well. One common criticism of prepayment in South African townships concerns how the technology contributes to depoliticising the relationship between state and citizens. Critics argue that prepayment allows the state to distance itself from the contestation and litigation that ensues from cut-offs when bills are not paid by households with conventional meters (Ruiters, 2007; Schnitzler, 2008; Heusden, 2009; Ruiters, 2009, 2011). The responsibility for managing the use of a collective service is said to shift from the state (or the utility provider in its place) to the consumer (Marvin & Guy, 1997). For some commentators this raises legal issues in countries like South Africa where access to basic urban services, including electricity, is deemed a core citizen right (Flynn & Chirwa, 2005). Through the processes of self-disconnections, the hand of the state or the utility provider becomes invisible, or socially privatised to individual households (Drakeford, 1998). As a result, critics argue, the lines of accountability for provision of basic utility services are made more tenuous. Opportunities for contestation are foreclosed, as any problems involved in consumption are now made the responsibility of the user alone.

These critiques call our attention to how access to collective urban services, such as the provision of electricity, water, sewage and waste management, are often considered a hallmark of citizenship and inclusion in modern societies and can, for that reason, be mobilised as a space of contestation (Holston, 2008). However, some critiques seem to simplify the diversity of arrangements for the service provision in place in many countries, whilst adopting a normative stance towards privatisation. They seem to reduce the interaction between utility users and providers to billing, a simplification derived from critics' over-attention to the economic aspects of the interaction. Moreover, contestation can happen at different junctures and take many forms, from overt protest to the micropolitics of everyday resistance (Scott, 1985; Bayat, 2004). Aspects like new connections to the grid, power outages, maintenance of meters, or upgrades to existing connections can also provide opportunities for contesting the relationship between electricity users and providers. These and other moments of the everyday consumption of electricity can provide opportunities for individuals to articulate their understanding of what being a 'citizen' means and what is expected (or not expected) of the abstract entity of the 'state'. As argued before, the prepayment meter (or the conventional meter for that matter) can be widely implicated in these public affairs. Following Domínguez Rubio and Fogué (2013), the visibility of the prepayment meter has the potential to make the politics of electricity consumption a 'matter of concern' instead of a 'matter of fact' – something that is open for public scrutiny instead of being taken for granted in urban life. Through everyday engagement with the meter, users may articulate a critical positionality vis-à-vis their participation in broader social relations. As a result, an expanded examination of the politics of prepayment outside of South Africa offers an opportunity to reflect on how the current trend towards the adoption of this technology may enable a specific politics of urban utility service provision in the context of poorly resourced urban areas, where the categories of 'state' and 'citizen' are in themselves fluid.

The Flows of Electric Power in Mozambique

The electrification of Mozambique is only now starting to pick up pace. In 2012, an estimated 22 per cent of the Mozambican population had access to electricity (EDM, 2014). However small this figure may seem, it stood at only four per cent a decade earlier (EDM, 2014). Moreover, access to electricity remains regionally uneven: a little more than half of the electricity available in 2012 was consumed in the South, particularly in Maputo (EDM, 2014). Perhaps this is unsurprising, considering that Mozambique remains a largely rural country by international standards.[2]

Prepayment has played a crucial role in the roll-out of electricity in Mozambique. The technology was first introduced in the Maputo area in 1995 with one pilot project connecting 500 households in the town of Matola and another one in 1996 connecting 5000 households in Maputo proper (Figure 6.1). The pilot projects came about as an attempt to address the state-owned electricity company EDM's cash-flow difficulties and a widespread problem with non-paying customers. At the time, Mozambique was coming out of nearly 16 years of civil war (1977–1992) and the country experienced a period of high-inflation and economic hardship. The Cahora Bassa dam, the country's most significant power generation plant, which is located on the Zambezi River in the Centre region, had been badly damaged during the civil war and it was producing little electricity at the time. Cahora Bassa came into operation in 1975, the year of independence, on the basis of a model of electricity export to South Africa negotiated earlier by the Portuguese. Because the electricity produced by the dam was transmitted directly to South Africa and there was no grid connecting the dam to Maputo, the deal allowed Mozambique to draw electricity from South Africa's grid at preferential prices to supply the city. When Cahora Bassa was compromised (from 1982 onwards) and the export deal could not be fulfilled, EDM had to buy electricity from South Africa at higher prices to supply Maputo. With the economy struggling to achieve a transition to Socialism at the doors of Apartheid South Africa (Hall & Young, 1997), EDM found it very difficult to generate enough revenue to match the higher electricity costs. The company ran a relatively long and inefficient billing cycle based on estimates of consumption with conventional meters. Contention with customers was rife and many refused to pay their bills.[3] In this context, officials at EDM and within the government decided to deploy prepayment as a solution to the cash-flow and non-payment problems. After two

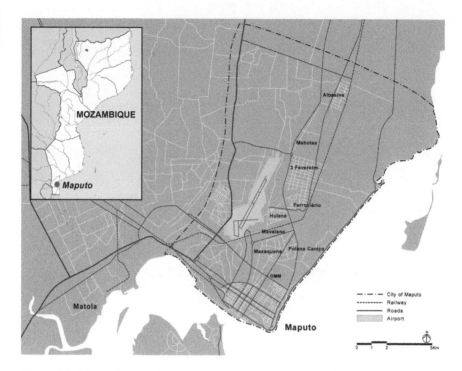

Figure 6.1 Schematic map of Maputo
Source: the author

successful pilot projects in the mid-1990s, EDM began substituting conventional meters for prepayment ones in all neighbourhoods of Maputo from the mid-2000s onwards, as well as in the rest of the country (Figure 6.2). In 2012, 90 per cent of EDM's clients in the city of Maputo were on a prepayment meter (EDM, 2014) and the company's target was to have all residential and small business customers fitted with a prepayment meter by the end of 2014. Prepayment also became the default electricity retail method countrywide: by 2012, 83 per cent of EDM's clients were using prepayment (EDM, 2014).

Policymakers and officials at EDM are quick to praise the success of the prepaid electricity system, known locally as *Credelec*, for transforming the electricity profile of Mozambique. Prepayment has allowed EDM to collect much needed revenue to slowly upgrade the incipient and obsolete grid left behind by the Portuguese, while keeping electricity tariffs artificially low. As a state-owned company and the single electricity distributor in Mozambique, EDM's business model is highly dependent on governmental politics. For some local commentators, the FRELIMO government, in power since independence, has used electrification as a political tool to win popular allegiance against opposition parties (namely RENAMO and MDM). While many accuse the government of long espousing

Figure 6.2 Maputo's prepayment meter
Source: the author

the neoliberal dictates of international donors and global capital, they also recognise how its policy for the energy sector has remained fluid and ambiguous
over time. For instance, while the prospection of offshore natural gas fields has
been essentially privatised, the government has resisted full liberalisation and
unbundling of the Mozambican electricity sector. In fact, informants close to the
Ministry of Energy and international donors note how the government has opted
politically to keep electricity tariffs unsustainably low over time, thus preventing the emergence of a competitive electricity market. Many see this strategy
as unsustainable in the long-term, especially as the internal demand for electricity keeps growing. The Cahora Bassa dam is currently operating near capacity
and EDM has limited capacity to generate more electricity on its own. Together
with the infrastructural dependency from South Africa on transmission, there is
an on-going anxiety in governmental circles about the country's energy sovereignty, matched by the appetite of Mozambican economic elites for developing
the energy sector. As a result, the Mozambican government set out an ambitious
plan of energy production and transmission mega-projects worth over USD$9
billion that is making many international donors nervous. Some of the projects
include the North-expansion of the Cahora Bassa dam, the new Mphanda Nkuwa

hydropower station downstream on the Zambezi River, a series of smaller hydro-power stations, natural gas- and coal-fired power stations, as well as a new energy transmission line connecting the Centre to the South regions, including Maputo (MacauHub, 2009). In spite of the government's commitment to rolling out these large infrastructure projects and other rural and urban electrification projects with the help of Chinese, Brazilian and Indian capital, Maputo continues to command most of the electricity consumption in the country.

As noted earlier, the energy profile of Maputo is somewhat different from that of the rest of the country. Unpublished EDM statistics indicate that about 97 per cent of the nearly 1.2 million estimated population of the city had access to an electricity connection in 2013.[4] Yet, having access to an electricity connection is not the same as having reliable access to quality electricity provision or consuming it liberally. In fact, there is a stark infrastructural difference between what locals call, on the one hand, the *cidade* (city) or *cidade de cimento* (city of cement) and, on the other hand, the *bairros* (neighbourhoods) and *subúrbios* (suburbs). The *cidade* is the consolidated colonial centre located on the waterfront and built to European standards. The *bairros* and the *subúrbios* are the neighbourhoods located outside of the *cidade* where Mozambicans were forced to locate during colonial times in dwellings made out of traditional materials (e.g. reed, wood and zinc). Nowadays, the *bairros* and *subúrbios* have expanded significantly outwards and most dwellings have been converted to cement block and/or brick construc-tions. This happened mostly through the endeavour of individual households and disparate interventions by governmental or international institutions, of varying degree of formalisation and planning. Jenkins (2012) estimates that residential development occupies nearly 30 per cent of the area of Maputo, 60 per cent of which is unplanned with the remaining area having some form of planned sub-division (both official and unofficial). Urban farming remains an important land use in Maputo, taking up an estimated 26 per cent of the area of the city, and functions as a complement to an urban economy that rests largely on informal activities – estimated at 65 per cent of the city's workforce (Jenkins, 2012). In practice this means that access to basic infrastructures – such as water, sewage and electricity – is limited and dependent on individual efforts to connect households to those services. A recent study of the socio-economic conditions of households in the neighbourhoods outside the *cidade* of Maputo casts light on the difficulty of this endeavour: nearly half of the households surveyed were considered poor or very poor, with only about 31 per cent of the workforce having some form of formal employment (e.g. as civil servants or in private businesses) (Andersen, 2012). Hence, even if households are able to connect themselves to basic services, that does not mean they can sustain their regular use.

In fact, access to electricity in Maputo's *bairros* and *subúrbios* has been the product mostly of individual endeavour in the context of the lifelong task of improving one's home (Bénard da Costa, 2011). During colonial times, access to electricity was largely restricted to the *cidade*. Outside of that, the population resorted usually to firewood, charcoal, candles and kerosene lamps for cooking and lighting (Rita-Ferreira, 1968; Mavhunga, 2013). After independence, there were programmes from the central government for the construction of pit latrines in the *bairros* through self-help community initiatives, but there was hardly any emphasis on residential electrification. Government concerns with the electricity sector remained at a high level, as exemplified by discussions around the control of Cahora Bassa dam (Isaacman & Isaacman, 2013). Individual households were left to their own devices when it came to 'putting energy in the house' (*pôr energia em casa*), the common Mozambican expression for wiring a dwelling. Accessing electricity seemed an inevitable and essential requirement of modern life. But connecting to the grid required both an engagement with challenging livelihood conditions and the enduring infrastructural deficits of Maputo's *bairros* and *sub-úrbios*. Wiring a dwelling could take a number of years depending on the ability of the household to save enough money for the electric installation and the extension of the grid from the nearest transformation post to the house. As in other aspects of everyday life, this could involve a wider network of extended family members, neighbours and friends, or members of the wider community (e.g. local state officials/politicians) (Bénard da Costa, 2011). The members of the network would lend money to each other, pool assets together, share services or facilitate access to EDM to speed up the connection process. It was not uncommon to pay some form of 'compensation' to EDM staff to speed up the process, overlook unfit installations or even to have to drive them from the headquarters to the neighbourhood to ensure that inspections and connections were carried out in a timely fashion. Often, individuals had to mimic a social order that reflected a form of inverse governmentality, 'a form of reason [by urban dwellers] that takes as its object the problem of governance', as Nielsen (2011: 329) put it. In other words, electricity users had to take on the responsibility of addressing the shortcomings of a state apparatus, here represented by the state-owned electricity company EDM, which was unable or unwilling to take the role of an effective and efficient provider of services to citizens.

Despite enduring concerns within EDM's leadership regarding these pervasive internal irregularities, the process of managing electricity users remained mired with many moments of uncertainty. Until the adoption of prepayment meters, electricity was billed on the basis of estimates, occasionally verified with on-site readings of the conventional meter by EDM staff. This was troublesome on many levels, both to the electricity provider and the users. A common story is that of EDM staff being bribed to not take meter readings. There are tales of certain neighbourhoods having an entrenched culture of non-payment, involving intricate schemes to avoid the visits of the meter readers. Many report having received electricity bills with unexpected sums to pay, either because the billing covered two or three months of service or because it accounted for an unexpected amount of electricity consumption. EDM would then be forced to disconnect non-paying households, deal with disgruntled customers, accusations of corruption and forego revenue. To electricity users, having to sort out incorrect billing or disconnections could be time-consuming, costly in terms of making trips to EDM and involve complicated paperwork and procedures that many could not follow. More importantly, it would force many into debt, meaning they had to pool resources once more to put energy *back* into the house. In sum, the conventional meter seemed to be a headache to all involved.

This overview highlights the complex politics of electrification and electricity provision in Mozambique, and Maputo in particular. It is difficult to pin the transition to prepayment as the result of a turn to neoliberalism per se (although it may be that too), when the politics of infrastructure remains entangled in the country's postcolonial legacy. For the most part, Mozambique's grid remains technically constrained by limited electricity generation and transmission infrastructure. In Maputo, the geographical limitations of the colonial grid have been slowly overcome over time as EDM responded to the demands of individual households outside the *cidade*. These demands were not necessarily vocalised publicly or openly. Instead, they were the result of a quiet appropriation of ordinary urban experiences by residents of the *bairros* and *suburbios* seeking an improved livelihood that would mimic the urban experience of the *cidade*. In this context, the state-society relationships embedded in the politics of urban energy infrastructure were at best fluid, if not ambiguous.

Practices of Prepaid Electricity in Maputo, Mozambique

In the context of pervading problems surrounding the conventional metering system, the population of Maputo's *bairros* and *suburbios* was generally welcoming of the shift to prepayment. It was not that electricity users were happy to be asked by EDM to pay for the new prepayment meters which replaced the old ones, to pay upfront for a low-quality service plagued by voltage fluctuations and power blackouts (especially if they had managed to have electricity for 'free' before) or to pay a per unit flat rate that was marginally more expensive than the conventional block tariff rates.[5] Rather, the much welcomed changes brought the possibility of avoiding the uncertainties inherent in the relationship with EDM as well as electricity-related debt, and being able to account for consumption according to disposable income.

In fact, many of the informants interviewed during fieldwork expressed their satisfaction with prepayment by highlighting the 'autonomy' and 'control' it gave them over spending and consumption. People consumed what they could afford, thus avoiding debt on unpaid bills estimated by EDM or spanning more than a month. In fact, being able to buy only small amounts of electricity credit depending on how much income the household managed to secure on any given day was a familiar practice in the acquisition of other goods (e.g. food or charcoal). This gave people a sense of empowerment over their uncertain lives, contributing to a sense of structural security over social life and making everyday livelihoods a bit more manageable (Adam & Groves, 2007). Most people acknowledged they made judicious decisions about daily electricity use, often disciplining themselves or the other members of the household. Such disciplining could involve unplugging certain appliances (e.g. refrigerators) to save electricity for other appliances (e.g. television). Or it could involve waiting to use an appliance collectively instead of for an individual alone (e.g. turn on the electric kettle only when everybody

wanted to have tea). Many people reported developing simple heuristics involving the meter's technology. Because the meter displays how many kilowatts-hour
(kWh) units of electricity are available for consumption, individuals learned to
associate a certain number of kWh with a specific amount of money and a given
number of days (or hours). Through their daily use of electricity for lighting or
different appliances, individuals developed an intuitive understanding of consumption patterns. This understanding did not require specific levels of technical
literacy about the meter's technology or the electricity system. It was practical
knowledge people developed from interaction with the prepayment meter. Many
households avoided running out of credit through these daily routines of 'disciplined autonomy' and it was surprising to hear few accounts of self-disconnections
among informants. However, admittedly, the disciplining most people undertook
was in itself a form of self-disconnection in which the prepayment meter played
the role of mediator in the organisation of social life.

The transition to prepayment was not without contestation, especially after the
mid-2000s when it started being rolled out to the whole city. Most of the protests
were about the cumbersomeness and inaccuracies of the prepayment sales system.
In principle, the prepayment system is simple: anyone from the household can go
to a vendor with however much money they want (or have), provide the number
of the prepayment meter installed in the dwelling and s/he will be given a 20-digit
numeric code to input into the meter's keypad to top it up. With the first purchase
of each month, the household is billed for two additional service fees: (1) the radio
and television fee (*taxa de radiodifusão*), collected by the central government; and
(2) the refuse fee (*taxa de lixo ou taxa de limpeza*), collected by the municipality. These services fees had been charged with the conventional electricity bill as
well, so it seemed unlikely the fees would constitute a problem. Yet, paying for
these service fees became a point of contention for many electricity users. With
the conventional system the service fees were 'matters of fact', buried as they
were in the lump sum of an estimated bill that many could not afford anyway. With
the prepayment system, the service fees became a 'visible' 'matter of concern'.
Every month electricity users were reminded that whatever money they had to buy
electricity top-ups, some of it was taken away to pay for services that, admittedly,
were deemed to be of poor quality (e.g. electricity supply), were not used (e.g. TV)
or were not even available to them (e.g. refuse collection).

Being reminded every month of the preposterousness of such payments in
the face of urban conditions in the *bairros* and *subúrbios* was enhanced by the
technical limitations of the initial client management sales system. The system
managed client accounts via an off-line terminal located at each vending shop,
either run by EDM or a private agent (e.g. pharmacies). Because there was no
communication between the systems of different shops, payment of service fees
were logged into the off-line terminal of the shop where the client had made the
first purchase of the month. This meant that individual customers always had to
buy electricity top-ups from the same shop or they would risk paying the service
fees more than once. Moreover, the slowness of the sales system meant customers faced long waiting times. It was only after mid-2012, when EDM upgraded

the prepayment retail system to an online server connecting all shops city-wide, expanded the network of vending shops through outsourcing and improved the sales interface (in late 2013), that problems with the service fees and long waits decreased. Since late 2013/early 2014, EDM made electricity top-ups available via mobile phones, ATMs, electronic wallet services and the Internet. While EDM and the private companies involved in this 'outsourcing' of sales (e.g. Vodacom, Millennium BIM bank or MPesa) have invested significantly in advertising the benefits of ubiquitous digital payments, the take-up has been slow. For the most part there are hardly any ATMs outside the *cidade,* and mobile connections or Internet access are not always reliable. Improvements in prepayment retail have been evident over the last two years, but this digital 'smart' urban energy system goes only so far to improve the urban livelihood conditions in Maputo. Practices of acquiring and consuming electricity remain seemingly fluid in the *bairros* and *suburbios* and in a liminal space between 'visible' 'matters of concern' and 'invisible' 'matters of fact'.

It is perhaps as a result of this fluidity that prepayment can become a terrain for articulating the nature of state-society relationships. While there were barely any public demonstrations in the face of these conditions, most conversations about prepayment evolved into discussions that shed light onto what people thought about different dimensions of the 'state' and their own condition as 'citizens'. Discontentment was always implicit in how customers treated vendors at the vending shops or in accusations directed towards EDM as being a 'bunch of thieves' or 'incompetent swindlers' by informants and in social media. A notable example of these accusations is a July 2014 post on Facebook by Azagaia, a famous Mozambican rapper, where prepayment was targeted in a direct critique of EDM. The rapper criticised EDM for suggesting that prepayment technology empowers users to control electricity consumption. Yet, as he argued, consumers were in fact powerless (in both senses) if EDM failed to make electricity available for consumption on the grid. In media outfits associated with the opposition, there are also occasional derogatory pieces on how EDM is disrespectful of its customers and how inherently disrespectful the service is to 'the Mozambican people'.[6] Moreover, conversations about how much the household spent on electricity per month often evolved into a discussion about the perceived inequality in Mozambican society. If asked about the quality of electricity supply, informants would elaborate on the uneven urban development conditions that, in their view, clearly distinguished the *cidade* from the *bairros* and *suburbios*. When talking about the long waits at a vending shop with security guards who enabled queue jumping if bribed, people would equate the guards' behaviour with the pervasive corruption and scandals traversing the political and economic arena. Even the contradictions of international policies such as the Millennium Development Goals could infiltrate conversations where residents weighed in on how controlling electricity consumption via prepayment meters allowed them to access what they perceived to be a better life – a life similar to that lived in the 'city of cement'. Overall, these examples suggest how everyday engagement with prepayment offered an opportunity for residents of the *bairros* and *suburbios* to articulate a critical positionality vis-à-vis that complex entity some call the 'government' – and which we tend analytically to designate as the 'state'.

Prepayment was thus one more terrain within which the relationship between 'state' and 'citizens' could be conceptualised and rationalised, where sense could be made of the difficult urban conditions outside the *cidade,* where the desires for a decent livelihood could be eked out amidst uncertain incomes and poor quality services. Political consciousness about the role of the government (and of its agents, especially EDM) remained prevalent among residents of Maputo's *bairros* and *subúrbios* in spite of, or because of, the transition to prepayment.

Conclusion

The on-going roll-out of prepaid electricity in Maputo, Mozambique, since the mid-1990s provides insights into urban politics on the grid and the socio-technical transition that is gathering pace everywhere on the African continent, as well as in other cities of the South. The perceived benefits of adopting prepayment meters accrued by utility service providers are spilling into other areas of the economy as well. In Nigeria, for example, local manufacturers of prepayment meters recently rallied against meter imports for taking over the market and stifling job creation in the sector (Metering International, 2014). The prospects of expanding access to basic urban infrastructure on the back of prepayment technology, common for other services such as pay-as-you-go mobile phones, fits with wider benign narratives of smart cities and even smart metering (although prepayment meters and smart meters are not the same). In other words, this seemingly and arguably unstoppable trend warrants attention from urban scholars not least for the potential that prepayment has for disregarding the needs of the urban poor, as many have pointed out already.

Yet, adopting an anti-prepayment stance that draws much of its argument from the perceived benefits of the 'modern infrastructural ideal' provided by a 'state' to the benefit of its 'citizens' seems to obscure how the delivery and consumption of basic utility services in cities like Maputo are deeply entangled in a complex and fluid postcolonial legacy. Findings from the Maputo case suggest that prepayment meters are reconfiguring how electricity is consumed in close connection with residents' pragmatic acknowledgement of the lived urban experiences in neighbourhoods where the 'modern infrastructural ideal' hardly ever materialised. In deploying a form of 'disciplined autonomy' mediated by the meter, households feel empowered to access a service they associate with a desired livelihood found mostly in the confines of Maputo's modern colonial 'city of cement'. The ubiquity and universality of prepayment, qualities of utility service prized within the 'modern infrastructure ideal', are also reconfigured with prepayment. It is possible to acquire electricity credit everywhere and nearly every household in Maputo seems to be connected to the grid. Yet, there is no guarantee there will be electricity to be consumed if generation does not expand in the near future; there is also no guarantee that individual households will have enough resources to consume electricity liberally if a large majority of the city's population continues to live on a hand-to-mouth basis. Perhaps because prepayment arguably makes 'visible' the challenges of consuming electricity due to the everyday practices it elicits, it has the potential to develop a particular politics of urban service provision and consumption that is only now starting to take shape. In the case of Maputo, such politics elicited in electricity users a complex and ambiguous reflection on their urban condition and the nature of wider state-society relationships. For the most part and until recently, this politics has remained largely unspoken, unlike the rather visible protests in neighbouring South Africa. As cities are being re-wired, or even wired anew, via prepayment electricity meters, this emerging techno-politics is bound to continue to shape the flows of power in the postcolonial African metropolis in ways that are historically constituted and geographically experienced.

Acknowledgements

This chapter draws on a working paper published online in 2013 by the Institute for Science, Innovation and Society, University of Oxford, UK, under the title *Everyday Practices of Prepaid Electricity in Maputo, Mozambique*. Research presented in this chapter has been conducted with the support of a grant from Oxford's John Fell Fund and support from the Research Fund of the Department for Continuing Education, University of Oxford, UK. For facilitating and making possible fieldwork in Maputo, I would like to thank Eng. Aires, Eng. Amilton, Eng. Bernardo, Pedro Coimbra, Luís Lage, Paul Jenkins, Sr. Manhiça, Aires Novela, Sr. Simbine and Maria Vitoria.

Notes

1 The market consultant firm Northeast Group, LLC, produced a ten-year market forecast report for prepayment technology in 15 Sub-Saharan countries in 2014. The fact that the report costs nearly USD$5,000 per enterprise client provides an indication (even if subjective) of the market interest in the technology.

2 The UN-HABITAT (2007) estimated the rural and urban populations in 2007 to be 64 and 36 per cent, respectively.

3 According to EDM internal statistics, nearly 16 per cent of the electricity available for consumption in 1995 was accounted as non-technical losses (i.e. pilfered electricity), while another 13 per cent was lost due to infrastructural deficiencies in transmission and distribution (UNDP & The World Bank, 1996).

4 Statistics provided by EDM during fieldwork in April 2014. These numbers should not be taken at face value, not least because the baseline of the city's population is somewhat fluid.

5 The block tariff system divides consumption into blocks of energy units charged at different, usually increasing, rates – e.g. in Mozambique, the cost of each kWh for domestic consumers is: from 0 to 300 kWh, $2.50 Meticais (USD¢8); from 301 to 500 kWh, $3.53 Meticais (USD¢11); above 501 kWh, $3.71 Meticais (USD¢12). The prepayment flat rate per kWh is $3.18 Meticais (USD¢10), but unlike post-payment clients, prepayment clients do not pay for a separate monthly electricity service fee *(taxa fixa)* of $85.35 Meticais (USD$2.7) (all 2014 prices, available at http://www.edm.co.mz/, accessed on 23 October 2014).

6 See, for instance, the opinion section 'Xiconhoca da Semana' of the newspaper *@Verdade* (http://www.verdade.co.mz/opiniao/xiconhoca).

References

Adam, B. and C. Groves (eds.), 2007. *Future Matters*. Leiden: Brill.

Akrich, M., 1992. The De-Scription of Technical Objects. In: W.E. Bijker and J. Law (eds.), *Shaping Technology/Building Society: Studies in Sociotechnical Change*. Cambridge: MIT Press, pp. 205–24.

Andersen, J.E., 2012. Home Space: Socio-Economic Study (Research Programme *Home Space in African Cities*, Funded by the Danish Research Council for Innovation 2009–2011). Copenhagen: The Royal Danish Academy of Fine Arts, School of Architecture, Department of Human Settlements.

Baptista, I., 2013. 'The Travels of Critiques of Neoliberalism: Urban Experiences from the "Borderlands"'. *Urban Geography* 34(5): 590–611.

Barry, A., 2001. *Political Machines: Governing a Technological Society*. London: The Athlone Press.

Bayat, A., 2004. Globalization and the Politics of the Informals in the Global South. In: A. Roy and N. Alsayyad (eds.), *Urban Informality: Transnational Perspectives from the Middle East, Latin America, and South Asia*. Lanham: Lexington Books, pp. 79–102.

Bénard Da Costa, A., 2011. Urban Transformation, Family Strategies and 'Home Space' Creation in the City of Maputo. In: S. Khan, P. Meneses and B. Bertelsen (eds.), *Dialogues with Mozambique. Interdisciplinary Reflections, Readings and Approaches on Mozambican Studies*. Leiden: Brill, 102–50.

Bijker, W.E., T.P. Hughes and T.J. Pinch (eds.), 2012. *The Social Construction of Technological Systems: New Directions in the Sociology and History of Technology (Anniversary Edition)*. Cambridge: The MIT Press.

Blundo, G. and P.-Y. Le Meur (eds.), 2009. *The Governance of Daily Life in Africa: Ethnographic Explorations of Public and Collective Services*. Leiden: Brill.

Briceño-Garmendia, C., A. Estache and N. Shafik, 2004. Infrastructure Services in Developing Countries: Access, Quality, Costs and Policy Reform (World Bank Policy Research Working Paper 3468). Washington, DC: The World Bank.

Casarin, A.A. and L. Nicollier, 2008. Prepaid Meters in Electricity. A Cost-Benefit Analysis (Working Paper IAE Business School). Buenos Aires: Austral University.

Casillas, C.E. and D.M. Kammen, 2010. 'The Energy-Poverty-Climate Nexus'. *Science* 330(6008): 1181–2.

Coutard, O., 2008. 'Placing Splintering Urbanism: Introduction'. *Geoforum* 39(6): 1815–20.

Domínguez Rubio, F. and U. Fogué, 2013. 'Technifying Public Space and Publicizing Infrastructures: Exploring New Urban Political Ecologies through the Square of General Vara del Rey'. *International Journal of Urban and Regional Research* 37(3): 1035–52.

Drakeford, M., 1998. 'Water Regulation and Pre-payment Meters'. *Journal of Law and Society* 25(4): 588–602.

EDM, 2014. *Sumário Estatístico/Statistical Summary 2012*. Maputo: Electricidade de Moçambique, E.P.

Estache, A., V. Foster and Q. Wodon, 2002. *Accounting for Poverty in Infrastructure Reform: Learning from Latin America's Experience*. Washington, DC: The World Bank.

Fankhauser, S. and S. Tepic, 2007. 'Can Poor Consumers Pay for Energy and Water? An Affordability Analysis for Transition Countries'. *Energy Policy* 35(2): 1038–49.

Ferguson, J., 2009. 'The Uses of Neoliberalism'. *Antipode* 41(S1): 166–84.

Flynn, S. and D.M. Chirwa, 2005. The Constitutional Implications of Commercializing Water in South Africa. In: D.A. McDonald and G. Ruiters (eds.), *The Age of Commodity: Water Privatization in Southern Africa.* London: Earthscan, pp. 59–75.

Graham, S. and S. Marvin, 2001. *Splintering Urbanism: Networked Infrastructures, Technological Mobilities, and the Urban Condition.* London: Routledge.

Guy, S. and S. Marvin, 1995. Pathways to 'Smarter' Utility Meters: The Socio-Technical Shaping of New Metering Technologies (Electronic Working Paper No. 23). Newcastle-upon-Tyne: Global Research Unit, School of Architecture, Planning and Landscape, University of Newcastle-upon-Tyne.

Hall, M. and T. Young, 1997. *Confronting Leviathan: Mozambique Since Independence.* London: C. Hurst & Co.

Harvey, E., 2005. Managing the Poor by Remote Control: Johannesburg's Experiments with Prepaid Water Meters. In: D.A. McDonald and G. Ruiters (eds.), *The Age of Commodity: Water Privatization in Southern Africa.* London: Earthscan, pp. 120–9.

Heusden, P.V., 2009. Discipline and the New 'Logic of Delivery': Prepaid Electricity in South Africa and Beyond. In: D.A. McDonald (ed.), *Electric Capitalism: Recolonising Africa on the Power Grid.* London and Cape Town: Earthscan and Humans Sciences Research Council Press, pp. 229–47.

Holston, J., 2008. *Insurgent Citizenship: Disjunctions of Democracy and Modernity in Brazil.* Princeton: Princeton University Press.

Holston, J. and A. Appadurai, 1996. 'Cities and Citizenship'. *Public Culture* 8(2): 187–204.

Isaacman, A.F. and B.S. Isaacman, 2013. *Dams, Displacement and the Delusion of Development: Cahora Bassa and Its Legacies in Mozambique, 1965–2007.* Athens: Ohio University Press.

Jaglin, S., 2008. 'Differentiating Networked Services in Cape Town: Echoes of Splintering Urbanism?'. *Geoforum* 39(6): 1897–906.

Jenkins, P., 2012. Home Space: Context Report (Research Programme *Home Space in African Cities,* Funded by the Danish Research Council for Innovation 2009–2011). Copenhagen: The Royal Danish Academy of Fine Arts, School of Architecture, Department of Human Settlements.

Kooy, M. and K. Bakker, 2008. 'Technologies of Government: Constituting Subjectivities, Spaces, and Infrastructures in Colonial and Contemporary Jakarta'. *International Journal of Urban and Regional Research* 32(2): 375–91.

Loftus, A., 2006. 'Reification and the Dictatorship of the Water Meter'. *Antipode* 38(5): 1023–45.

Macauhub, 2009. Mozambique: Mozambican Energy Projects Cost US$9 Billion. [Online]. 17 March. Available from: http://www.macauhub.com.mo/en/2009/03/17/6733/. [Accessed: 8 May 2013].

Marres, N., 2012. *Material Participation: Technology, the Environment and Everyday Publics.* Basingstoke: Palgrave Macmillan.

Marvin, S. and S. Guy, 1997. 'Consuming Water: Evolving Strategies of Water Management in Britain'. *Journal of Urban Technology* 4(3): 21–45.

Mavhunga, C.C., 2013. 'Cidades Esfumaçadas': Energy and the Rural-urban Connection in Mozambique'. *Public Culture* 25(2): 261–71.

McDonald, D.A. (ed.), 2009. *Electric Capitalism: Recolonising Africa on the Power Grid.* London and Cape Town: Earthscan and Humans Sciences Research Council Press.

McDonald, D.A. and G. Ruiters (eds.), 2005. *The Age of Commodity: Water Privatization in Southern Africa.* London: Earthscan.

McInnes, P., 2005. Entrenching Inequalities: The Impact of Corporatization on Water Injustices in Pretoria. In: D.A. McDonald and G. Ruiters (eds.), *The Age of Commodity: Water Privatization in Southern Africa*. London: Earthscan, pp. 99–119.

Metering International, 2014. Nigeria's Manufacturers Rally Against Imports of Prepayment Meters. [Online]. 3 September. Available from: http://www.metering.com/nigerias-manufacturers-rally-against-imports-of-prepayment-meters/. [Accessed: 26 October 2014).

Naidoo, P., 2007. 'Struggles Around the Commodification of Daily Life in South Africa'. *Review of African Political Economy* 34(111): 57–66.

Nielsen, M., 2011. 'Inverse Governmentality: The Paradoxical Production of Peri-urban Planning in Maputo, Mozambique'. *Critique of Anthropology* 31(4): 329–58.

Pachauri, S., A. Mueller, A. Kemmler and D. Spreng, 2004. 'On Measuring Energy Poverty in Indian Households'. *World Development* 32(12): 2083–104.

Parnell, S. and J. Robinson, 2012. '(Re)Theorizing Cities from the Global South: Looking Beyond Neoliberalism'. *Urban Geography* 33(4): 593–617.

Plancq-Tournadre, M., 2004. 'Services d'eau et d'électricité au Cap, ou comment la sortie de l'apartheid fabrique des débranchés'. *Flux* 56–57: 13–26.

Prins, G., I. Galiana, C. Green, R. Grundman, A. Korhola, F. Laird, T. Nordhaus, R. Pielke Jr., S. Rayner, D. Sarewitz, M. Shellenberger, N. Stehr, and H. Tezuko, 2010. *The Hartwell Paper: A New Direction for Climate Policy After the Crash of 2009*. Oxford and London: Institute for Science, Innovation and Society, University of Oxford and the London School of Economics and Political Science Mackinder Programme.

Rita-Ferreira, A., 1968. 'Os Africanos de Lourenço Marques'. *Separata de Memórias do Instituto de Investigação Científica de Moçambique* 1967/1968: 93–491.

Robinson, J., 2006. *Ordinary Cities: Between Modernity and Development*. London: Routledge.

Robinson, J., 2011. 'The Travels of Urban Neoliberalism: Taking Stock of the Internationalization of Urban Theory (2010 Urban Geography Plenary Lecture)'. *Urban Geography* 32(8): 1087–109.

Ruiters, G., 2007. 'Contradictions in Municipal Services in Contemporary South Africa: Disciplinary Commodification and Self-disconnections'. *Critical Social Policy* 27: 487–508.

Ruiters, G., 2009. Free Basic Electricity in South Africa: A Strategy for Helping or Containing the Poor? In: D.A. McDonald (ed.), *Electric Capitalism: Recolonising Africa on the Power Grid*. London and Cape Town: Earthscan and Humans Sciences Research Council Press, pp. 248–63.

Ruiters, G., 2011. 'Developing or Managing the Poor: The Complexities and Contradictions of Free Basic Electricity in South Africa (2000–2006)'. *Africa Development* 36(1): 119–42.

Sagar, A.D., 2005. 'Alleviating Energy Poverty for the World's Poor'. *Energy Policy* 33(11): 1367–72.

Schnitzler, A.V., 2008. 'Citizenship Prepaid: Water, Calculability, and Techno-politics in South Africa'. *Journal of Southern African Studies* 34(4): 899–917.

Schnitzler, A.V., 2013. 'Traveling Technologies: Infrastructure, Ethical Regimes, and the Materiality of Politics in South Africa'. *Cultural Anthropology* 28(4): 670–93.

Scott, J.C., 1985. *Weapons of the Weak: Everyday Forms of Resistance*. New Haven: Yale University Press.

Scott, J.C., 1998. *Seeing like a State: How Certain Schemes to Improve the Human Condition Have Failed*. New Haven: Yale University Press.

Simone, A., 2000. *On Informality & Considerations for Policy (Dark Roast Occasional Paper Series n.º 3)*. Kenilworth: Isandla Institute.

Simone, A., 2004. 'People as Infrastructure: Intersecting Fragments in Johannesburg'. *Public Culture* 16(3): 407–29.

Tewari, D.D. and T. Shah, 2003. 'An Assessment of South African Prepaid Electricity Experiment, Lessons Learned, and their Policy Implications for Developing Countries'. *Energy Policy* 31(9): 911–27.

UN-HABITAT, 2007. *Perfil do Sector Urbano em Moçamnique*. Nairobi: The United Nations Human Settlements Programme.

UNDP and The World Bank, 1996. *Mozambique Electricity Tariffs Study, Report No. 181/96 – ESMAP, Energy Sector Management Assistance Programme*. Washington, DC: The United Nations Development Programme and the World Bank.

Part III

Social Movements and Protest in the Electric City

7 Berlin

Cooperative Power and the Transformation of Citizens' Roles in Energy Decision-Making

Arwen Colell and Luise Neumann-Cosel

Introduction

Widespread calls for a transition towards a renewable and sustainable energy system in Berlin have resulted in new types of social and political vitality for local distribution grids – a new role for what can be termed the 'last mile' of the power grid. Local distribution grids are now the key infrastructural domains for balancing demand and supply from decentralised and renewable energy generation, playing a crucial role in determining the direction and speed of the *energiewende,* the German word for a local energy transition (Diekmann et al., 2007; Kenkmann & Timpe, 2012). Showing the 'fundamentally networked character of urbanism' (Graham & Marvin, 2001: 5), these developments highlight the essential nature of electricity infrastructures within the configuration of both urban economies and politics (Marvin et al., 1999). The implications of the resulting urban grids transcend path dependencies that shape present socio-technical systems (Tarr & Dupuy, 1988), exploring new technical, financial and managerial configurations to mediate and operationalise a shift towards sustainable energy systems at the community level (Kaika & Swyngedouw, 2000; Geels & Kemp, 2007; Monstadt, 2007).

This chapter focuses on the *energiewende* of Berlin, Germany. Specifically, it traces the social movement led by BürgerEnergie Berlin (in English, Citizen Energy Berlin), a cooperative formed for the purpose of building social and financial capital towards putting the city's grid under the direct control of citizens. With a single share set at €100, the 1,000+ members of the cooperative raised over €10 million before presenting their bid for a partnership with the city of Berlin in June 2014 to establish joint grid operations (BürgerEnergie Berlin, 2014a). The choice of grid operator in Berlin, to be decided in summer 2015 (as this chapter was submitted for publication), opens a window of opportunity for infrastructural change and thus a shift in the policies and politics of urban electricity and sustainability more broadly. BürgerEnergie Berlin challenges well-established dichotomies around public and private management of public services, proposing a new form of infrastructural relations between municipal authorities and citizens. The case of Berlin highlights the interconnectedness of a transition in energy system and urban governance, which may inform a more meaningful debate on the merits and limits of the (re)municipalisation of utility services. Furthermore, it highlights the importance of greater clarity in understanding 'civic engagement' and 'community energy' architectures (Hoffmann & High-Pippert, 2005) around infrastructural politics.

The drastic political ruptures experienced by the city in the twentieth century, and their effects on the power grid, illustrate the close relationship between socio-political urban change and energy policy. As early as the 1930s almost 70 per cent of Berlin households were connected to the electricity grid, whilst household consumption was aggressively encouraged. As Moss (2014: 1433) argues, 'Berlin had come to symbolise the networked city of Europe'. Typical of a centralised and vertically integrated market, generation, retail and grid operations were managed by the city's own energy utility, BEWAG. Today, with 3.5 million inhabitants, Berlin has the country's single largest urban distribution grid. It covers approximately 2.3 million private and corporate electricity connections. BEWAG was privatised in the course of market liberalisation in the 1990s, when retail markets were opened and unbundled to separate energy generation, distribution and retail. German customers may now choose from over 1,000 different energy providers (IWR, 2013). However, regionally dominant energy utilities remain the main actors in the electricity sector throughout Germany, with grid operations and basic electricity services typically provided by subsidiaries of the same company (Becker & Templin, 2013; Berlo & Wagner, 2013).

The liberalisation of the energy market was established via concessions – awarded by municipal authorities – assigning the management and operations of local power grids to private sector players. These concessions typically last between 5 and 20 years, a temporality that opens the city to infrastructural and, by extension, socio-political change. Given its introduction in the late 1990s, this system has resulted in a nationwide expiration of almost all grid concessions between 2015 and 2018. As a result, a debate around the ownership and management of the public electricity grid has reopened in many communities.

In examining the implications of the cooperative sector's endeavour to recast public-private partnerships in infrastructure management, this chapter explores a three-fold reconstruction of electricity infrastructures: technical, financial and institutional. It argues that the establishment of citizen-owned cooperative structures for the purpose of running the city's electricity grid – in this case, operating in partnership with the municipality and responding to specific political, social and moral drivers (Berger, 2009) – may counter sceptics of what Skopcol (2003) refers to as civic culture: a sustained civic involvement in the management of public affairs (see also Bosso, 2003). More importantly, these cooperative efforts open new pathways for public participation beyond the simple indirect representation offered by local voting structures, and provides an impetus to wider decentralisation approaches within urban governance and infrastructure development. We argue that cooperatives present new opportunities for municipalities to engage in partnerships with their citizens, increasing the participation and self-efficacy of their constituency.

The chapter is divided in four sections. The following section briefly outlines different paradigms associated with decision-making within the public and private management of infrastructures. It provides the frame for the second section, an examination of key challenges in the realisation of the *energiewende* as a socio-technical endeavour. This section focuses on the political, economic and technical issues involved in decision-making over ownership and management. Such an analysis highlights why the local power grid has become a focus point of citizens' initiatives seeking a different role vis-à-vis their municipality in shaping their infrastructures and surroundings. The third section explores the case of BürgerEnergie Berlin, a citizen-owned cooperative seeking to establish a qualified model of public-private partnerships via joint operations with the municipality for the management of the local power grid. The fourth and final section discusses the reconstruction of infrastructures of decision-making and financing, as implied by the cooperative's approach. This final section examines the implications of the model proposed by BürgerEnergie Berlin for ongoing debates around the public vs. private management of urban utilities. The chapter concludes by revisiting the merits of the cooperative sector in building community energy architectures and promoting civic culture through increasing political, social and value-driven engagement.

Negotiating Infrastructures of Decision-Making for Networked Utilities

In a debate still largely dominated by the paradigm of private vs. municipal management, Berlin provides a relevant case study for illustrating the struggle and potential of citizens claiming more direct access to ownership and management of networked infrastructures. Throughout Germany calls for communal management of public infrastructures are becoming increasingly popular, counteracting the trend of the last few decades that favours the private management of infrastructures (Spohr, 2009). In the case of electricity, private grid operators are fighting hard to prevent such developments, a counter-trend playing out in the context of a variety of other potential difficulties standing in the way of (re)municipalisation. Public management does not necessarily strengthen local decision-making processes and often has difficulties in realising specific infrastructure projects through public income alone. Critics have called into question the merits of focusing the debate exclusively on public vs. private ownership and management. In line with this critique, a case needs to be made for a more profound definition and transformation of the overarching principles and objectives associated with the management of public utilities (Naumann & Lederer, 2010). If the local energy transition is to succeed, a more profound transformation is required, wherein ownership is considered to be a key component of the political, social and moral engagements with networked infrastructures (Hoffmann & High-Pippert, 2005). Reconstructing public infrastructure to implement the *energiewende* must go beyond a return to old paradigms of public management in redefining the roles of energy infrastructure as well as its owners and operators.

Communal vs. Private Management of Grid Infrastructures in Germany and Berlin

Growing municipal engagement in grid operations in Germany stems from an awareness of the strategic potential of the urban grid towards local value creation and tax revenues. This approach, concerned also with issues of climate change mitigation and the *energiewende*, is further encouraged by growing public support for the public management of key services, including water, health care, public transport and housing (Röber, 2009). Such reinvigoration of municipal management stems from the end of concessions lasting 15 to 20 years, forcing communal administrations to re-evaluate long-term plans, rather than pointing to a mass exodus of private interest in utility management.

However, the trend towards the municipalisation of public services faces a series of practical challenges, including asymmetrical regulatory conditions for market competition on local distribution grids. While in the past privatisation was encouraged by public entities giving away their assets, its reversal is now less popular with current (private) grid operators. Current concession holders can frequently exploit an ambiguous legal framework and draw political power from their strong economic position vis-à-vis municipal administrations to prevent communal takeover, ensuring their continued position in the market (Becker & Templin, 2013; Berlo & Wagner, 2013). Private grid operators' tactics in securing their control over local infrastructures include refusing disclosure of relevant information concerning existing infrastructure – a feature that blocks possibilities for competition – and incentives or sanctions to the local administration. Incentives include sponsorships, neighbourhood support or commitments towards job creation, whereas the threat of sanctions may comprise a loss of local jobs through a change of company headquarters (Reimer, 2002). Close ties between local administrations and private energy utilities within administrative boards or advisory committees further serve to blur the lines between private and public service providers as well as build pressure to ensure cooperation (Carini, 2013).

Municipalities choosing new concession holders frequently face former grid operators refusing or impeding commissioning and the unbundling of grid structures by withholding payment of license fees or even refusing negotiations altogether. Far from being exceptions, these tactics are 'a nationwide phenomenon' (Becker & Templin, 2013: 10), discouraging local authorities from entering into partnerships with new stakeholders. However, while highlighting the monopolistic position of municipal administrators (as the sole provider of concessions), the strong market position of the operator (as the sole provider of relevant information on the infrastructure at hand) is not adequately recognised by relevant authorities within the energy sector, particularly the *Bundesnetzagentur* (BNetzA), the federal authority regulating electricity markets. Becker and Templin (2013) point out that the misconduct of private energy utilities holding concessions goes against the German Energy Industry Act (or EnWG), as well as – for example in the case of threatened withdrawal of sponsorships – the law against restraints on competition. A coalition of different groups is calling for more stringent application of the law, most notably by the BNetzA and the *Bundeskartellamt* (BKAmt), the federal-level independent competition authority whose task is to protect competition in Germany (Becker & Templin, 2013: 18; Bund der Energieverbraucher, 2013; Krischer et al., 2013).

Civil society organisations such as BürgerEnergie Berlin go beyond this criticism of the inherent favouring of private over public management, questioning the sincerity of public programmes for increased civil society participation (BürgerEnergie Berlin, 2013a, 2013b), as well as the efficiency of municipalities in challenging established grid operators and indeed the foreclosing of a concessionary decision in their favour (BürgerEnergieBerlin, 2014a). A return to public management of infrastructure systems would, at best, result in indirect representation of local communities with limited opportunities for accessing transparent information and taking into account citizen voices in related decision-making processes (Monstadt, 2007). More importantly, however, associations such as BürgerEnergie Berlin claim transparency in decision-making processes and opportunities for direct interaction and participation of citizens (BürgerEnergie Berlin, 2013b, 2014a). It comes as an unsurprising consequence that citizens' attention is now turning to the underlying infrastructures mediating the system transition, specifically the local distribution grid.

Urban Lifelines – Why the Local Distribution Grid?

Twenty years after market liberalisation, municipalities and citizens alike are revisiting questions around the efficient and appropriate management of public networked infrastructures. Following the Chernobyl disaster of 1986, and seeking to step away from nuclear energy, a parents' initiative in the small town of Schönau, in the Black Forest, founded the first citizen-owned cooperative grid operator in Germany. In the context of a pre-liberalised energy market, a takeover of low-voltage grid operations presented their only chance of switching to renewable energy supplies. Although exemplary in its successful establishment of a citizen-owned grid operator and renewable energy utility (EWS, 2012), this citizen takeover of grid operations did not, at the time, spark many further attempts. Since then the liberalisation of the energy markets altered the electricity generation, retail and distribution landscape, transforming relationships between public and private actors towards infrastructure provision. Today, the decisions faced by municipalities reconsidering their options, in the context of expiring concessions, demand three important questions. What is at stake for the realisation of the *energiewende* as a socio-technical transformation? What is at stake for the local economy? And, lastly, what is at stake politically, beyond the mere issue of public vs. private management?

The Local Power Grid and the Energiewende

A shift to renewable energy results in an increasingly decentralised generation network together with the need for greater balance in supply and demand on a local scale. Beyond the technical implications of this bi-directional network and the need to put in place systems tailored to a fluctuating generating cycle, a key challenge is the re-alignment of the social interests and structures that traditionally operate the conventional energy system. In developing a local electricity grid with the ability to receive renewable loads, the socio-technical challenges are manifold. These range from a new structure of pricing mechanisms, efficiency incentives and possible links to behavioural change affecting consumers, through to the rise of interconnected 'smart' systems and the need to plan and manage a new geography of decentralised generation and 'smart' consumption. The socio-technical changes required to advance the *energiewende* point to the significance of institutionalising the transition processes beyond its technical domains, articulating the process with the demands of an increasingly environmentally concerned public (Fenn, 1999). An understanding of networked infrastructures as disconnected, black-boxed and seemingly naturalised entities, whose technical workings are indifferent to their socio-political surroundings (Graham & Marvin, 2001) must thus give way to new socio-economic configurations of power utilities (Deudny & Flavin, 1983; Morris, 2001; Hoffmann & High-Pippert, 2005). Viewed from this perspective, the *energiewende* is not a technical issue, but predominantly a social and political one.

Economics of Grid Management

As a regional monopoly, low-voltage grid operations in Germany are closely regulated on EU, national and municipal levels. Grid operators charge a consumer fee per kWh, whilst paying a license fee to municipal authorities. Consumer fees finance investments by the operator, covering development as well as maintenance and technical improvements. This fee is determined by federal authorities, depending on both the condition of the existing infrastructure and the regionally specific context of grid operations. Commercially, this framework renders the grid as both a safe and profitable asset, delivering considerable returns to operators. The local electricity context is thus shaped by both political drivers within municipal authorities and the organisational structure and financial interests of grid operators. Such conditions also shape how electricity relates to local economies and the aim of increasing the share of renewable energy across different locations (Kenkmann & Timpe, 2012).

Across Germany, a growing public trend is favouring the municipalisation of grid operations (Simunkova, 2013: 201), demanding profits to be tied to local value creation whilst questioning the performance and appropriateness of private management. Academic analyses underscore the potential gains for municipalities as well as positive management outcomes (Libbe, 2011). Such demands challenge federal laws that focus on energy security and availability of supply, and underscore the importance of a grid operator's individual commitment to realising the *energiewende* (Bechtel et al., 2013). The choice of grid operator significantly shapes local network design (including its technical ability to increase the share of renewables), with important implications for the future economic, technological and environmental trajectory of local electricity services. Tying in with these demands, citizens and small private investors are increasingly seeking local investment opportunities creating the conditions for the *energiewende* (Deuse, 2014), and these are being recognised as attractive financial investments for both public and private actors. In July 2013, the federal government introduced a 'citizens' dividend' to support investments in transmission capacities. As a means of speeding up grid improvement whilst also addressing public reluctance for over-land transmission lines, citizens were encouraged to invest in the electricity sector with a guaranteed five per cent return on investment (BMU, 2013). This *Bürgerdividende* programme, introduced as a means of increasing public participation in energy infrastructures and canvassing support for the *energiewende*, was soon criticised for being a highly speculative financial transaction (Ankenbrand & Kremer, 2013). Rather than opening pathways into closely regulated and comparatively low-risk financial investments, the *Bürgerdividende* programme turned citizens into creditors of energy utilities, redirecting financial risks to citizens. Such increased public participation via financial instruments did not, however, result in an increased voice in decision-making processes regarding the development and implementation of energy projects (Ankenbrand & Kremer, 2013; Doemens, 2013).

Governing the Local Power Grid: Politicising Naturalised Structures

Sustainable energy systems require meaningful forms of political, economic and social citizen participation, involving citizens in the ownership and management of networked infrastructures. We argue that in studying the governance of urban energy infrastructures it is necessary to go beyond issues of large-scale project management, energy security or public vs. private management (Hodge & Greve, 2005; Hughes & Ranjan, 2013). Sustainable electricity systems require informed and reflective customers. In Germany, such change inevitably challenges the underlying premises that have shaped infrastructure management since World War II. It means highlighting the ways by which material infrastructures are linked to (rather than disconnected from) their socio-political context. In some cases, it requires the establishment of new political configurations for grid operations; specifically, the rejection of management strategies designed to discourage political, social or economic engagement of citizens with their local energy system, together with an abandonment of the rhetoric of technocratic infrastructure management (Hoffmann & High-Pippert, 2005; Swyngedouw, 2007). Such an approach brings questions of social and civic engagement to the forefront of infrastructural (re)configuration. Beyond mourning the loss of 'civic culture' (Skopcol, 2003) or the conceptual stretching of the term 'civic' when conceptualising forms of engagement (Berger, 2012), scholars have built numerous arguments for more robust modes of participation (Williamson, 1997; Berger, 2009, 2012). Much of this draws on Barber's notion of 'strong democracy' (1984: 132), which importantly points to 'politics (citizenship) as a way of living: a fact of one's life, an expected element of it, a prominent and natural role in the same manner as that of parent or worker' (Prugh, Costanza & Daly, 2000: 112).

Networked infrastructures such as the local distribution grid may present ideal focal points for building a form of civic culture, providing platforms for 'sustained attention to issues to create a sense of community that transcends identity based upon a narrow reading of self-interest' (Hoffmann & High-Pippert, 2005: 91). Such new infrastructural relations could form a key element for the advancement of a wider democratic governance of cities. The institutionalisation of more democratic forms of electricity provision and infrastructure management must reflect the premise that community energy must not be 'conflated with local ownership' (Bolinger, 2004; Hoffmann & High-Pippert, 2005: 392). They also need nurturing, not only of the 'capacity for self-reliance but for citizenship' (Morris, 2001: 7). BürgerEnergie Berlin, as a citizens' initiative for the establishment of a cooperative form of infrastructure management, seeks to establish new partnerships with the Berlin municipality towards the advancement of sustainable energy in the city, with a view to changing citizen participation and involvement in local energy infrastructures.

BürgerEnergie Berlin: Citizens Are Claiming the Local Distribution Grid

Over the past two decades, the governance of electricity grids in German municipalities has been shaped by liberalisation processes (Héritier, 2002). In the 1990s Berlin took a leading role in the privatisation of public assets, partly due to the city's financial crisis post-reunification (Krätke, 2004: 513; Beveridge, 2012). It is argued that the city failed to install appropriate monitoring and evaluation mechanisms for the emerging hybrid service provision systems. This has resulted in poor performance on regional innovation and environmental modernisation, alongside limited economic benefits for the city (Monstadt, 2005). Unsurprisingly, a growing discontent with private utility ownership and an increase in awareness of the potential financial benefits to be gained from grid operations has given rise to a renaissance of ideas around community management of grid operations. Such developments are in line with larger national trends around the reversal of private utility management, most prominent in the water sector.

BürgerEnergie Berlin, which roughly translates to 'Citizens' Energy Berlin', has taken advantage of this window of opportunity, challenging simple dichotomies of public vs. private management by introducing a citizen-owned cooperative as a partner for the municipality for the purpose of electricity service provision. Its key objectives include strengthening democratic decision-making processes around infrastructure management, supporting local value creation by promoting direct financial ties to citizens and reinvesting profits in local projects. Its purpose is the development of a cooperative grid operator, joining public and private models through citizen-owned management partners (BürgerEnergie Berlin, 2014b). In the BürgerEnergie Berlin model, an injection of financial capital via citizen-owned shares, directly connecting cooperative members' political voice with the actions of the grid operator, is the key to controlling the grid (Figure 7.1). This provides a small but stable financial profit for cooperative members, enabled by regulated revenues whilst supporting local value creation.

Figure 7.1 Rephrasing Berlin's marketing slogan
Source: Sebastian Tuschel/BürgerEnergie Berlin

The cooperative sector is challenging the dominant role of Germany's private sector within energy utilities. Despite cooperatives being responsible for a high proportion of renewable energy generation – with more than 130,000 members (90 per cent of them individual citizens) investing approximately €1.2 billion in the *energiewende* in 2013 (DGRV, 2013) – established private energy utilities still control much of the grid. Their strong market position provides them with political leverage vis-à-vis regional and national regulatory authorities in the sector (Becker & Templin, 2013; Berlo & Wagner, 2013). Yet, an increase in decentralised generation is stretching traditional transmission and distribution networks beyond breaking point, and predicating changes in established infrastructure management strategies (Diekmann et al., 2007; Kenkmann & Timpe, 2012).

As a direct response to this energy landscape, BürgerEnergie Berlin argues that the inclusion of a citizen-owned cooperative in grid operations provides opportunities for urban dwellers to participate more directly in the policy and regulatory decision-making processes, at both local and federal levels. By democratising ownership, the cooperative's proposal goes one step beyond the municipalisation of electricity infrastructure. It reconnects Berlin's citizens and their networked infrastructures whilst striving to create a political climate favouring decentralised structures of decision-making and local value creation. It does this by giving power to citizens regarding decisions concerning their immediate electricity infrastructure. BürgerEnergie Berlin aims for an increase in public awareness of and participation in energy issues, through a model that also frames the electricity grid as a secure and sustainable financial asset for citizen investors. By giving citizens direct access to Germany's single largest distribution grid, the idea of the *energiewende* becomes a tangible objective for those not usually engaged in energy and environmental governance (Erb, 2012).

When presenting their formal bid for cooperation to the Berlin Senate in June 2014 (Figure 7.2), the cooperative had succeeded in signing up over 2,000 members and collecting over €10 million reserved for buying shares in the grid (BürgerEnergie Berlin, 2014b). More importantly, the initiative had generated a public debate around the electricity grid as an essential urban good, putting pressure on political parties at the federal Länder and municipal levels to publically commit to support direct citizens' participation in grid operations (Bartsch, 2013). The difficulties experienced by the cooperative through the formation and bid submission process resulted in increased media attention regarding questions of civil engagement in the city's grid management on a national and international level (Goldenberg, 2013; Grefe, 2013; Stonington, 2013; Subramanian, 2014). By October 2014, when the city's negotiations for assigning the concession were postponed, the cooperative had already been officially invited by the Berlin Senate to talks concerning their bid for the construction of cooperative grid operations (BürgerEnergie Berlin, 2014b).

BürgerEnergie Berlin points to an important political shift towards reimagining the role of urban and municipal stakeholders in the governance of energy systems.

Figure 7.2 Members of BürgerEnergie Berlin call on the city mayor in front of the Berlin Senate, after submitting the official bid

Source: Sebastian Tuschel/BürgerEnergie Berlin

This is a novel process in that it mobilises decentralised and polycentric structures that 'blend scales and engage multiple stakeholder groups' towards a new politics of the grid (Goldthau, 2014: 5; see also Ostrom, 2010; Brown & Sovacool, 2011; Schönberger, 2013).

Reconstructing Urban Infrastructure Beyond Public vs. Private

Key to the cooperative's project is the belief that a mere shift back to public service provision may not entail the desired changes in infrastructure management and its socio-technical embedding. New forms of social entrepreneurship or public-private partnerships may instead point the way towards reconstructing electricity infrastructure in more sustainable ways whilst advancing more inclusive forms of value creation, increased investment and democratic decision-making within service provision (Röber, 2009). Citizen initiatives play a key role in reshaping the politics of the grid, through calling for legislation that facilitates the emergence of communal forms of service providers and by establishing innovative private companies themselves. In the two largest cities of Germany, Hamburg and Berlin, citizens successfully lobbied municipalities to hold referendums on creating new public utilities for the provision of public services – in the case of Hamburg, electricity distribution. Although in both cases municipal authorities opposed the creations of public utilities, the successful referendum of Hamburg forced the municipality to take action (Unser Hamburg-Unser Netz, 2013). In Berlin the referendum failed to achieve the required 625,000 votes by just a few thousand, making it hard to deny the broad alliance that was forming in favour of public engagement in energy utilities (BürgerEnergie Berlin, 2013a). Beyond expressing a mere desire for state-owned utilities, the referendums illustrate public discontent with current forms of infrastructural governance. This underscores the need for a redefinition of the roles of public and private organisations as well as the forms of involvement of citizen groups. In the case of grid management, since the choice of concession holder remains governed by processes established under energy industry laws and regulations, a referendum operates only as a political signal in favour of a change of management paradigm. While the democratic legitimacy of publicly owned utility operators can be strengthened through a referendum with positive results, the vote's transformative potential, beyond reinstating public management, remains limited.

BürgerEnergie Berlin illustrates a different path for civil society organisations seeking to influence grid operations, by directly bidding for the concession of electricity distribution and pressuring the municipality for a transformation in infrastructure governance. A transformative reconfiguration of electricity grids implies the reconciliation of representative and participatory democracy, impacting structures of decision-making whilst establishing novel economic models. Attempts to bridge this gap find their expression in citizen initiatives seeking a direct participation in local decision-making beyond the established channels of representative democracy. This does not necessarily contradict models of public management, but instead presents an innovative and essentially pragmatic mode of reconciling representation and participation in established democracies (Pogrebinschi, 2013). Traditionally, political systems in established democracies are based on representative channels with limited opportunities for direct participation. The federal *Bürgerdividende* programme exemplifies the difficulties of introducing meaningful modes of participation, which consider political and economic elements within utility provision, beyond financial interaction (Jungbluth, 2013). By building public-private partnerships in infrastructure management, however, both financial participation and control may be strengthened. It is this combination of communal and cooperative structures that BürgerEnergie Berlin seeks to establish in Berlin.

Cooperatives as an associational form of ownership are well established within the German entrepreneurial tradition, playing a historical role in supporting local development and offering financial participation and direct representation (Münkner, 2015). Cooperatives feature prominently in the banking and housing sectors, but also in agriculture and, more recently, account for the majority share in citizen-financed renewable energy generation projects (DGRV, 2013). Each member of the cooperative has one voice, independent of his or her financial share, and benefits from the profit according to the scale of the investment made. While hedging against a single-interest takeover, this ownership form still incentivises sizeable financial investments. Cooperatives are publicly registered and monitored by accredited federal institutions regarding their economic performance, and have acquired the reputation of being reliable and stable investments (DGRV & AGENTUR FÜR ERNEUERBARE ENERGIEN, 2013). Through a cooperative model, the communal management of the electricity grid provides a degree of indirect representation of a broad base of community members. It offers strategic advantages, for instance towards advancing political agendas such as the local *energiewende,* and presents economic advantages through profits and tax revenues. A cooperative provides members with options for participating in strategic decisions, as well as offering a means of control vis-à-vis public players and an opportunity for broadening the base for profit appropriation.

Taking Back the City: But Who Can, and for Whom?

The role of citizen-owned cooperatives in creating political power for members whilst advancing greater economic inclusion in the process of transitioning networked infrastructures remains in flux. The implications of energy cooperatives for community energy architectures must take into account the considerable success of German cooperatives in the electricity generation sector, where their continued growth and relevance illustrates both their viability and attractiveness to citizens. Cooperatives present an opportunity to tackle some of the challenges associated with building long-term community energy engagement on diverse scales. Apart from the Schönau electricity rebels, who operated under very different conditions from those shaping the infrastructural landscape today, cooperatives seeking to take over or at least take part in distribution grid operations remain in the early stages of development.

We argue that the success of BürgerEnergie Berlin in taking forward a civil initiative towards a novel form of sustainable infrastructural politics relies in channelling what Berger (2009) defines as political, social and moral engagement. While 'civic engagement' is frequently highlighted as a crucial element of healthy democratic societies (Putnam, 1995; Schudson, 1996; Barber, 1998), its 'conceptual stretching' has rendered the term in many ways useless through a lack of any coherence, or even utility in operationalisation (Gerring, 1999 in Berger, 2009). In contemplating 'which kinds of engagement [. . .] make democracy work, and how they might be promoted', Berger (2009: 336) introduces the notions of political, social and moral engagement as a more meaningful and valuable alternative to civic engagement. He suggests that political engagement encompasses any 'attentive activity directly involving the polity' and designed to have an impact on government activity in any form (Berger, 2009: 341). This involves 'most of the activity that we normally associate with political participation or citizenship', but differs from social engagement in that it does not require multiple parties (Berger, 2009: 341). In contrast, social engagement entails forms of associational involvement that may or may not address political objectives (Berger, 2009). Finally, moral engagement addresses 'attention to, and activity in support of, moral code, moral reasoning or moral principles' (Berger, 2009: 342). Political or social engagement will not always contain or be directed towards morally desirable outcomes (Rosenblum, 1998).

Energy cooperatives such as BürgerEnergie Berlin encompass all three forms of engagement. Focusing on Berlin, the cooperative was established with a clear strategy and political agenda to shape public debate. The joint engagement of members of the cooperative and campaigners resulted in a form of social engagement. Firstly in their association as individuals, secondly as a collective within the cooperative, and thirdly in their engagement with the wider community through campaigns such as the promotion of a local referendum on energy issues (BürgerEnergie Berlin, 2013a). Finally, a clear purpose towards building a sustainable energy system that promotes local climate change mitigation and environmental protection points to a form of moral engagement. The degree of political and moral engagement within energy cooperatives may vary depending on a cooperative's broader agenda beyond its immediate cause, whilst social engagement as a joint activity in associational structures is generally assumed to be high. Cooperatives promoting an increase in renewable energy generation capacities may not immediately set a political agenda for themselves. Nevertheless, their embedding in an overarching discourse of an energy transition actively advocated by citizens has promoted their perception as politically and morally active agents of systemic transition (DGRV, 2013). At the same time, the interconnectedness of political, social and moral engagement with the forms of economic responsibility embedded within the cooperative model point to a new role for citizens in community development.

The type of political, social and moral engagement found in energy cooperatives is frequently illustrated through the notion of community energy architectures (Hoffmann & High-Pippert, 2005: 399). Cooperatives may be conceptualised as 'thin' or 'thick' operationalisations of community energy concepts. A thin interpretation of cooperatives as community energy agents sees the cooperative as a professionalisation and externalisation of energy interests. Such a conceptualisation resonates with ideas of stealth democracy (Hibbing & Theiss-Morse, 2002), wherein the many may avert political, social or moral engagement with energy transitions by invoking and supporting engagement of the few. This is exemplified by cooperatives with large membership numbers but often low attendance to annual general assemblies (Geschwandtner & Wieg, 2001). Membership and financial investment then serves to satisfy the need for political, social and moral engagement. In the case of cooperation through a public-private partnership, such as the one envisaged by BürgerEnergie Berlin, members are also presented with what a stealth democracy model assumes to be even more important than actual participation (Hibbing & Theiss-Morse, 2002: 2): the 'opportunity to participate if they should ever be motivated to do so, and they want to know that the power of their elected representatives could be checked by their own political power' (Hoffmann & High-Pippert, 2005: 389).

Public-private partnerships between municipalities and their citizens, achieved via a cooperative model, may thus present new forums of discourse and deliberation for community participation and an envisioned transition towards more sustainable forms of living (Grant, 2003). Cooperatives, in a thick conceptualisation of community energy, are devised as agents of a strong democracy (Barber, 1984). This position advocating strong democracy also cautions that participation could not result in full-time engagement of citizens. Advocates of 'a more robust conceptualisation of community energy [. . .] guided by [notions of] strong democracy' (Hoffmann & High-Pippert, 2005: 399) point to the difficulties associated with the institutionalisation of community involvement, as well as the limits of engendering broad public debate on complex and often technical issues. Whilst cooperatives provide a forum for institutionalised and long-term commitment with a strong community focus, their regulation around conditions of membership, their statutes and actions provides close control of their performance and actions. The Schönau and Berlin cooperatives actively engage community debate with experts and laymen in monthly and annual publications and conferences held for members and the interested public (BürgerEnergie Berlin, 2015; EWS, 2015). Their role as watchdogs of municipalities and private energy utilities creates a new infrastructural actor promoting transparency and providing monitoring.

At the same time, cooperatives face the challenge of balancing community activism and entrepreneurial professionalisation. As companies, especially in the closely regulated sector of grid operations, they compete with large and established market forces. Yet, they differ greatly to these competitors. As community actors, they seek to provide a forum for participation and information for members as well as the wider public. For BürgerEnergie Berlin, this included professionalising a still largely volunteer led team whilst ensuring an institutionalisation of community activism that remains closely connected to the grassroots. Taking back a city's infrastructure, in this sense, refers also to reconstructing operative structures between operating in a market and wider socio-environmental objectives.

Conclusion

The promise and potential of BürgerEnergie Berlin highlights the importance of a renewed debate on the role of the local scale in the configuration of urban electricity systems and the development of a different type of infrastructural politics. It calls into question traditional relationships between private utilities, public administrations and energy consumers through a reconfiguration of consumers into citizens. Cooperatives such as BürgerEnergie Berlin are moving from working towards the *energiewende* through increased generation capacity, to a more profound role in shaping energy infrastructures and future transition pathways. This entails a new forum of political, social and moral engagement for citizens in shaping their community, and merges political and economic citizenship in enabling meaningful civic participation in infrastructure investments and their revenues. The task of BürgerEnergie Berlin, subsequently, lies in its application for grid operations and the respective financial and technical foundations of its work. This is predicated on building public awareness regarding the relevance of utilities and related services for communities and the efficacy of citizen engagement on a local level. Independent of the municipality's choice of grid operator, expected for the summer of 2015, the evaluation of cooperative engagement in grid operations already indicates a new forum for civil participation and public awareness regarding networked infrastructures. The Berlin cooperative succeeded in creating public debate not only on the social, economic and environmental direction of grid management but also on its underlying paradigms concerning ownership and decision-making.

BürgerEnergie Berlin is an expression of increased public interest in networked infrastructures. Citizens increasingly demand a less politically detached notion of infrastructure management, claiming voice and inclusion in related decision-making processes and revenue streams. Such demands, in arguing for more informed and direct means of participation, transcend national and international trends towards the privatisation of utility services. The case of Berlin highlights the cooperative as a tool of building political, social and moral engagements with networked infrastructures that institutionalise these in new socio-technical architectures of electricity. Beyond questions around the possibility of cooperative-led takeovers of infrastructure operations within current regulatory frameworks, this experience holds important repercussions for the recurring question of how to design and operationalise transitions towards more sustainable energy systems. Energy cooperatives may focus on generation, distribution or retail, but their embedding of civic engagement may foster a new wave of renewable and community electricity projects in diverse settings.

The experience of BürgerEnergie Berlin, however, also illustrates the many difficulties associated with transforming the scale from where infrastructures are managed. Applying a cooperative approach, typically realised in smaller structures characterised by strong communal ties, to a project of considerable technical and financial dimensions is not without challenges. The pioneering Schönau 'power rebels' owe much of their success to a strong community engagement thanks to face-to-face interactions. BürgerEnergie Berlin, instead, is facing the trade-off of a highly visible large project and an attendant structure less amenable to meaningful mobilisation and engagement. Just as difficult as the need to increase membership and foster public identification with the cooperative's aims, the transformation of forms of ownership within well-established utility models challenge financial and management structures. This may be bridged by overcoming the information deficit that the general public still has in relation to networked infrastructures and their management structures (BürgerEnergie Berlin, 2012). Beyond the cooperative's financial capacity, successfully taking over the distribution grid depends on its political strength, something that is achieved exclusively through a meaningful engagement of citizens. Therefore, the ability of citizens to identify with the object and aims of the cooperative is imperative. Ultimately, the long-term success of the cooperative depends on the interest of the many, rather than merely the work of the few engaged in the bid application process and entrepreneurial tasks involved. As a monitoring actor but also in case of direct participation in grid management, the cooperative itself must preserve and reinvigorate its democratic decision-making processes in order to fulfil its aims of improving citizen control of the grid. BürgerEnergie Berlin draws attention to the electricity grid in cities as a political site crucial to mediating the debate around sustainability on a local level (Agyeman, 2005; Bickerstaff et al., 2009).

References

Agyeman, J., 2005. *Sustainable Communities and the Challenge of Environmental Justice.* New York: New York University Press.

Ankenbrand, H. and D. Kremer, 2013. Alarm, Bürgerdividende! *Frankfurter Allgemeine Zeitung.* 20 July. [Online]. Available from http://www.faz.net/aktuell/wirtschaft/wirtschaftspolitik/ausbau-der-stromtrassen-alarm-buergerdividende-12289120-p2.html?printPagedArticle=true#pageIndex_2. [Accessed: 17 January 2015].

Barber, B.R., 1984. *Strong Democracy: Participatory Politics for a New Age.* Berkeley: University of California Press.

Barber, B.R., 1998. *A Place for Us: How to Make Society Civil and Democracy Strong.* New York: Hill and Wang.

Bartsch, D., 2013. Energienetze in Bürgerhand. *DIE WOCHE.* [Online]. Available from http://www.linksfraktion.de/kolumne/energienetze-in-buergerhand/ [Accessed: 17 January 2015]

Bechtel, A., C. Niehoerster and A. Nolde, 2013. *Kurzkonzept. Der Beitrag des Netzbetreibers zur Energiewende.* [Online]. Available from http://www.buerger-energie-berlin.de/wp-content/uploads/Kurzkonzept_Energiewende.pdf [Accessed: 11 January 2015].

Becker, P. and Templin, W., 2013. Missbräuchliches Verhalten von Netzbetreibern bei Konzessionierungsverfahren und Netzübernahmen nach §§ 30, 32 EnWG. *Zeitschrift für Neues Energierecht ZNER* 1: 10–18.

Berger, B., 2009. 'Political Theory, Political Science and the End of Civic Engagement'. *Perspective on Politics* 7(2): 335–50.

Berger, B., 2012. *Attention Deficit Democracy: The Paradox of Civic Engagement.* Princeton: Princeton University Press.

Berlo, K. and O. Wagner, 2013. Auslaufende Konzessionsverträge für Stromnetze – Strategien überregionaler Energieversorgungsunternehmen zur Besitzstandswahrung auf Verteilnetzebene. [Online]. Available from http://epub.wupperinst.org/files/4856/4856_Konzessionsvertraege.pdf. [Accessed: 11 January 2015].

Beveridge, R., 2012. *A Politics of Inevitability: The Privatization of the Berlin Water Company, the Global City Discourse and Governance in 1990s Berlin.* Wiesbaden: Verlag für Sozialwissenschaften.

Bickerstaff, K., H. Bulkeley and J. Painter, 2009. 'Justice, Nature and the City'. *International Journal of Urban and Regional Research* 33(3): 591–600.

BMU, 2013. Rösler und Altmaier: Bürgerdividende soll Netzausbau beschleunigen und breitere Akzeptanz für die Energiewende schaffen. [Online]. 13 July. Federal Ministry of the Environment, Nature Conservation and Nuclear Safety. Available from http://www.bmwi.de/DE/Presse/pressemitteilungen,did=583574.html. [Accessed: 18 June 2014].

Bolinger, M., 2004. *A Survey of State Support for Community Wind Power Development.* Berkeley: Lawrence Berkeley National Laboratory and the Clean States Energy Alliance.

Bosso, C.J., 2003. 'Rethinking the Concept of Membership in National Advocacy Organisations'. *The Policy Studies Journal* 31(3): 397–411.

Brown, M.A. and B.K. Sovacool, 2011. *Climate Change and Global Energy Security: Technology and Policy Options.* Cambridge, MA: MIT Press.

BUND DER ENERGIEVERBRAUCHER, 2013. Rekommunalisierung mit Hindernissen. *Press Statement.* [Online]. Available from http://www.energieverbraucher.de/de/Buero-Verkehr/Kommunen/Konzessions—vertraege_1673/ContentDetail_13592/. [Accessed: 17 January 2015].

BürgerEnergie Berlin, 2012. *Allparteien-Konsens: Berliner Fraktionen wollen Bürgergenossenschaft am Stromnetz beteiligen.* [Online] 4 November. Available from http://www.buerger-energie-berlin.de/presse. [Accessed: 18 June 2014].

BürgerEnergie Berlin, 2013a. *Koalition: Bürgermitsprache unerwünscht.* [Online]. 29 August. Available from http://www.buerger-energie-berlin.de/presse [Accessed: 17 June 2014].

BürgerEnergie Berlin, 2013b. *Wowereit kann sich vor dem Volksentscheid retten, aber nicht vor den Berlinern.* [Online]. 4 November. Available from http://www.buerger-energie-berlin.de/presse. [Accessed: 17 June 2014].

BürgerEnergie Berlin, 2014a. *Bürgergenossenschaft unter letzten drei Stromnetzbewerbern.* [Online]. 16 June. Available from www.buerger-energie-berlin.de/presse. [Accessed: 17 June 2014].

BürgerEnergie Berlin, 2014b. *Kurz und gut.* Newsletter to Members and Followers of BürgerEnergie Berlin [Email newsletter]. 15 July.

BürgerEnergie Berlin, 2015. *Konzessionsvertrag ausgelaufen – und nun?* [Online]. 8 January. Available from www.buerger-energie-berlin.de/blog [Accessed: 9 January 2015].

Carini, M., 2013. Auch Vattenfall steht zur Wahl. *TAZ.* [Online]. 21 September. Available from http://www.taz.de/!124085/. [Accessed: 17 June 2014].

Deudney, D. and C. Flavin, 1983. *Renewable Energy. The Power to Choose.* New York: W.W. Norton & Company.

Deuse, K., 2014. Erfolg mit nachhaltiger Geldpolitik. *Deutsche Welle.* [Online]. 14 June. Available from http://www.dw.de/erfolg-mit-nachhaltiger-geldpolitik/a-17389132. [Accessed: 17 June 2014].

DGRV, 2013. *Großer Zulauf bei Energiegenossenschaften: Bereits 130.00 Mitglieder engagieren sich.* [Online]. 24 July. Deutscher Genossenschafts- und Raiffeisenverband. Available from http://www.dgrv.de/webde.nsf/272e312c8017e736c1256e31005c

edff/5f450be165a66e4dc1257c1d004f7b51/$FILE/Pressemitteilung%2024.7.2013.pdf. [Accessed: 17 June 2014].

DGRV and AGENTUR FÜR ERNEUERBARE ENERGIEN, 2013. *Energiegenossenschaften. Bürger, Kommunen und lokale Wirtschaft in guter Gesellschaft.* [Online]. Available from http://www.dgrv.de/webde.nsf/272e312c8017e736c1256e31005cedff/dbc 67594818be17dc1257c1d004fc167/$FILE/Brosch%C3%BCre%20Energiegenossen schaften.pdf. [Accessed: 17 June 2014].

Diekmann, J., U. Leprich and H.J. Ziesing, 2007. *Regulierung der Stromnetze in Deutschland. Ökonomische Anreize für Effizienz und Qualität einer zukunftsfähigen Netzinfrastruktur.* Edition der Hans-Böckler-Stiftung 187.

Doemens, K., 2013. Bürgerdividende mit Risiken. *Berliner Zeitung.* [Online]. 22 July. Available from http://www.berliner-zeitung.de/wirtschaft/anwohner-renditebeteiligung-buergerdividende-mit-risiken,10808230,23787586.html. [Accessed: 17 June 2014].

ELEKTRIZITÄTSWERKE SCHÖNAU EWS, 2012. *Wir werden immer mehr.* [Online]. Available from http://www.ews-schoenau.de/ews/layer-ews-in-zahlen/kundenanstieg. html?print=1. [Accessed: 17 June 2014].

ELEKTRIZITÄTSWERKE SCHÖNAU EWS, 2015. *Stromseminar 2015: Programm.* [Online]. Available from http://www.ews-schoenau.de/mitmachen/stromseminar.html [Accessed: 18 January 2015].

Erb, S., 2012. Bürger wollen an die Leitung. *TAZ.* [Online]. 25 April. Available at: http://www.taz.de/Energie-I-/!92214/. [Accessed: 17 June 2014].

Fenn, P., 1999. Without Community Choice, No Consumer Choice. *Local Power News.* [Online]. Available from www.local.org. [Accessed: 13 April 2014].

Geels, F.W. and R. Kemp, 2007. 'Dynamics in Socio-Technical Systems: Typology of Change Processes and Contrasting Sase Studies'. *Technology in Society* 29(4): 441–55.

Gerring, J., 1999. 'What Makes a Concept Good?' *Polity* 31(3): 357–93.

Geschwandtner, M. and A. Wieg, 2001. 'Generalversammlung via Internet'. *GenossenschaftsVerband* 2001(4): 22.

Goldenberg, S., 2013. Berliners' Co-op Aims to Take Over and Run Electricity Grid of City. *The Guardian.* [Online]. 28 August. Available from http://www.theguardian.com/world/2013/aug/28/berliners-co-op-aims-run-electricity [Accessed: 7 May 2015].

Goldthau, A., 2014. 'Rethinking the Governance of Energy Infrastructure: Scale, Decentralization and Polycentrism'. *Energy Research & Social Science* 1(3): 134–40.

Graham, S. and S. Marvin, 2001. *Splintering Urbanism: Networked Infrastructures, Technological Mobilities and the Urban Condition.* London: Routledge.

Grant, J.A., 2003. *Community, Democracy, and the Environment: Learning to Share the Future.* Lanham, MD: Rowman and Littlefield.

Grefe, C., 2013. Berlins Stromrebellen. *DIE ZEIT.* [Online]. 6 June. Available from http://www.zeit.de/2013/23/stromnetz-berlin/komplettansicht [Accessed: 1 May 2014].

Héritier, A., 2002. 'Public-interest Services Revisited'. *Journal of European Public Policy* 9(6): 995–1019.

Hibbing, J.R. and E. Theiss-Morse, 2002. *Stealth Democracy: America's Beliefs about How the Government Should Work.* New York: Cambridge University Press.

Hodge, G.A. and C. Greve, 2005. *Challenge of Public–Private Partnerships: Learning from International Experience.* Cheltenham: Edward Elgar.

Hoffmann, S. and A. High-Pippert, 2005. 'Community Energy: A Social Architecture for an Alternative Energy Future'. *Bulletin of Science, Technology and Society* 25(5): 387–401.

Hughes, L. and A. Ranjan, 2013. 'Event-related Stresses in Energy Systems and their Effects on Energy Security'. *Energy* 59: 413–21.

IWR, 2013. *Deutschland hat so viele Strom- und Gasanbieter wie noch nie.* Internationales Wirtschaftsforum Erneuerbare Energien. [Online]. Available from http://www.iwr.de/news.php?id=24072. [Accessed: 1 May 2014].

Jungbluth, R., 2013. Und wer zahlt für die Extrawurst? Altmaier und Rösler wollen Stromnetzgegner schmieren. *Die ZEIT* No. 29, 11 July 2013.

Kaika, M. and E. Swyngedouw, 2000. 'Fetishizing the Modern City: The Phantasmagoria of Urban Technological Networks'. *International Journal of Urban and Regional Research* 24(1): 120–38.

Kenkmann, T. and C. TIMPE, 2012. *Dezentral, ressourcenschonend, effizient: Bausteine einer zukunftsfähigen Energieversorgung.* Freiburg: Öko-Institut e.V.

Krätke, S., 2004. 'City of Talents? Berlin's Regional Economy, Socio-Spatial Fabric and "Worst Practice" Urban Governance'. *International Journal of Urban and Regional Research* 28(3): 511–29.

Krischer, O., J.-H. Fell, B. Hasselmann, E. Ebner, B. Herlitzius, B. Höhn, F. Ostendorff, D. Steiner, D. Wagner and V. Wilms, 2013. Kleine Anfrage: Entwicklung der Anzahl der Gerichtsverfahren bei der Konzessionsvergabe für Strom- und Gasnetze. *Deutscher Bundestag, 17. Wahlperiode. Drucksache 17/14438.* [Online]. Available from http://dip21.bundestag.de/dip21/btd/17/144/1714438.pdf. [Accessed: 18 June 2014].

Libbe, J., 2011. Rekommunalisierung als Trend und Chance für Kommunen? *Deutsches Institut für Urbanistik, difu Berichte 3/2011.* [Online]. Available from http://www.difu.de/publikationen/difu-berichte-32011/rekommunalisierung-als-trend-und-chance-fuer-kommunen.html. [Accessed: 1 May 2014].

Marvin, S., S. Graham and S. Guy, 1999. 'Cities, Regions and Privatised Utilities'. *Progress in Planning* 51(2): 91–165.

Monstadt, J., 2005. *Die Modernisierung der Stromversorgung. Regionale Energie- und Klimapolitik im Liberalisierungs- und Privatisierungsprozess.* Wiesbaden: Verlag für Sozialwissenschaften.

Monstadt, J., 2007. 'Urban Governance and the Transition of Energy Systems: Institutional Change and Shifting Energy and Climate Policies in Berlin'. *International Journal of Urban and Regional Research* 31(2): 326–43.

Morris, D., 2001. *Seeing the Light: Regaining Control of our Electricity System.* Minneapolis, MN: Institute for Local Self-Reliance.

Moss, T., 2014. 'Socio-Technical Change and the Politics of Urban Infrastructure: Managing Energy in Berlin between Dictatorship and Democracy'. *Urban Studies* 51(7): 1432–48.

Münkner, H.-H., 2015. Die Internationale Bedeutung von Genossenschaften. *Akademie der Genossenschaften.* [Online]. Available from http://www.adgonline.de/adg_online/Internationales-Verbaende/News/2013/internationale-bedeutung-genossenschaften/ [Accessed: 16 February 2015].

Naumann, M. and K. Lederer, 2010. 'Öffentlich, weil es besser ist? Politische Gemeinwohlbestimmung als Voraussetzung einer erfolgreichen Kommunalwirtschaft'. *Berliner Debatte Initial* 2010(4): 105–16.

Ostrom, E., 2010. 'Beyond Markets and States: Polycentric Governance of Complex Economic Systems'. *The American Economic Review* 100(3): 641–72.

Pogrebinschi, T., 2013. 'The Squared Circle of Participatory Democracy: Scaling Up Deliberation to the National Level'. *Critical Policy Studies* 7(3): 219–41.

Putnam, R.D., 1995. 'Bowling Alone: America's Declining Social Capital'. *Journal of Democracy* 6(1): 65–78.

Prugh, T., R. Costanza and H. Daly, 2000. *The Local Politics of Global Sustainability.* Washington, DC: Island Press.

Reimer, N., 2002. Der Osten in schwedischer Hand. *TAZ.* [Online]. Available from http://www.taz.de/1/archiv/archiv-start/?ressort=wu&dig=2002%2F02%2F20%2Fa0085&cHash=f2bc7e4cce9438e077b7bba7fe6105c4 [Accessed: 17 June 2014].

Röber, M., 2009. 'Privatisierung adé? Rekommunalisierung öffentlicher Dienstleistungen im Lichte des Public Managements'. *Verwaltung & Management* 5: 227–40.

Rosenblum, N., 1998. Compelled Associations. In: A. Gutmann (ed.), *Freedom of Association.* Princeton: Princeton University Press, pp. 75–108.

Schönberger, P., 2013. Municipalities as Key Actors of German Renewable Energy Governance: An Analysis of Opportunities, Obstacles, and Multi-level Influences. *Wuppertal Papers, No. 186.* [Online]. Available from http://nbn-resolving.de/urn:nbn:de:bsz:wup4-opus-46766. [Accessed: 1 May 2014].

Schudson, M., 1996. 'What If Civic Life Didn't Die?' *American Prospect* 25 (March–April): 17–27.

Simunkova, T., 2013. Ick will meinen Strom zurück. *Der Stern.* [Online]. 29 May. Available from http://www.stern.de/politik/deutschland/trend-rekommunalisierung-ick-will-meinen-strom-zurueck-2017345.html. [Accessed: 17 June 2014]

Skopcol, T., 2003. *Diminished Democracy: From Membership to Management in American Civic Life.* Norman: University of Oklahoma Press.

Spohr, G., 2009. 'Rekommunalisierung der Energieversorgung. Revival der Stadtwerke'. *Planerin* 6: 16–18.

Stonington, J., 2013. Big Energy Battle – An Unlikely Effort to Buy the Berlin Power Grid. *SPIEGEL.* [Online]. Available from http://www.spiegel.de/international/business/an-unlikely-effort-to-buy-the-berlin-power-grid-a-886426.html [Accessed: 1 May 2014].

Subramanian, C., 2014. Women Take Aim at Berlin's Power. *The Daily Beast.* [Online]. 4 March. Available from http://www.thedailybeast.com/witw/articles/2014/03/04/two-women-campaign-to-put-berlin-s-electricity-grid-in-the-people-s-hands.html [Accessed: 14 June 2014].

Swyngedouw, E., 2007. Sustainability and Post-Political and Post-Democratic Populism: Reclaiming Democracy? In: Van Christaller tot Wallerstein (ed.), *Liber Amicorum.* Zelzate, Belgium: Nautulus Academic Books, 99–118.

Tarr, J. and G. Dupuy (eds.), 1988. *Technology and the Rise of the Networked City in Europe and America.* Philadelphia: Temple University Press.

Unser Hamburg–Unser Netz, 2013. *Volksentscheid gewonnen.* [Online]. 23 September. Available from http://unser-netz-hamburg.de/volksentscheid-gewonnen-ein-gutes-ergebnis-fur-hamburg/. [Accessed: 10 April 2014].

Williamson, T., 1997. *What Comes Next: Proposals for a Different Society.* Washington, DC: National Center for Economic and Security Alternatives.

8 Beirut
Metropolis of Darkness: The Politics of Urban Electricity Grids

Eric Verdeil

Introduction

Despite massive investment in infrastructure reconstruction, Beirut has never fully recovered 24/7 provision of electricity after the civil war (1975–90). Since 2006, partly as a result of Israeli bombings, infrastructure decay has worsened the situation. In addition, an atmosphere of political bickering in the country has delayed the implementation of new projects. Lighting for the majority of households (93 per cent) is supplied through a publicly owned transmission and distribution grid run by the state utility Electricité du Liban (EDL) (Central Administration of Statistics, 2008). However, on average, electricity is supplied for only half of the day, severely affecting the daily life of the city's nearly two million inhabitants, who are forced to cope with extended periods of darkness and generalised power failures. As a result, across the majority of the population and through all facets of urban life, electricity flows have been substantially reorganised around alternative informal electrical grids running on privately owned small-scale generators. These emerging flows, crafted to cope with the pervasive reality of power cuts, build new infrastructural, social and political dynamics in the city. The multiple facets of electricity and its shortages have become the subject of heightened controversies in the political arena, fuelling various forms of political mobilisation and challenging government policies in several crucial areas. The domains at stake transcend the very materiality of power cuts, including the management of utility workers in an era of neoliberalisation as well as struggles for power and control over the privately owned and informally established electricity generators now powering the city. Unfolding protests centred on electricity have taken to the street, threatening the nation's political order, the unity of the country and, seemingly, its very existence (Figure 8.1).

Researching electricity crises in Lebanon poses unique dilemmas given their embeddedness in everyday life. In Beirut, like in many other cities of the global South, everyday infrastructural interruptions – such as power outages, traffic congestion and road blocks – represent a 'new normality' (McFarlane, 2010). Yet, as McFarlane (2010: 133) suggests, such 'everyday forms of interruption have received relatively little attention in urban studies'. Beirut's electricity crisis, as an everyday force, has become a major factor in the (re)production of the uneven

Figure 8.1 A smelting Lebanon, lit by candle
Source: Abdelhalim Hammoud (used with permission)

geography of the city. Through this, it produces a new urban politics, the subject of this chapter. As McFarlane and Rutherford argue, 'the politics underpinning urban infrastructural transformation are rarely more evident or visible than in times of crisis or rupture' (2008: 6). The electricity crisis is openly contested in the 'back alleyways' of the city (Bayat, 2009: 11) as much as in Parliament, on social networks through cyber-mobilisation, and on the streets through public demonstrations. This chapter thus questions how these new forms of electric 'politics' reframe broader 'political' domains in the city (Swyngedouw, 2011). In doing so, the chapter is in line with approaches which, within urban and infrastructure studies, displace the centrality of regulatory reform and the governance of the sector (Graham & Marvin, 2001; Coutard, 2008). Rather, its concern lies in an analysis of the impacts of infrastructure policies on the social fabric of cities (Monstadt, 2009; Jaglin & Verdeil, 2013; Rutherford & Coutard, 2014), unpacking issues of energy justice (Bickerstaff et al., 2013), particularly at an urban scale.

Consistent with other academic accounts of the politics of urban infrastructures (Heynen et al., 2006; McFarlane & Rutherford, 2008; Graham, 2010), the uneven geography of electricity supply in Beirut reproduces existing social and political hierarchies. It favours the city of Beirut and its richer population over that of its suburbs whilst also reinforcing the political and sectarian lines that divide the country. The main objective of this chapter is to analyse the forms of mobilisation and politicisation around electricity issues in Beirut, recognising the centrality of the material and spatial dimensions of the infrastructure network in these emerging forms of social protest. Here infrastructure acts to redistribute the agencies of power in the city, contesting existing social and political hierarchies and prompting the emergence of new, alternative but precarious, configurations of power. Such a novel configuration can be further understood through Timothy Mitchell's notion of a 'power from within the energy system' (Mitchell, 2011: 12), in this case via the city's electricity grid. This involves reflecting on the grid's political geography and examining how urban politics shape grid politics. The main body of the chapter is divided into four sections. The first, drawing on the work of Mitchell (2011) and Cupples (2011), proposes a framework for analysing the politicisation of infrastructure in cities. This is followed by a geographical analysis of the social and political context of the Lebanese electricity crisis, highlighting the prevailing uneven access to electrical power and the resulting street protests that, with little success, have occurred during the last ten years. The third section examines the different forms of protest and dissent that, by drawing on its material and relational properties, are entangled in the electricity grid. Specifically, two forms of grid-embedded resistance are examined, both representing what has been termed as 'the end of the [electricity] line': the ability of electricity workers to disrupt the system through strikes and the capacity of users to establish forms of passive resistance through meter tampering. The fourth section scrutinises the new social and power relationships built around alternative and informal electricity networks based on private generators. Finally, the conclusion elaborates a typology based on various combinations of grid and urban politics.

The chapter's empirical material consists of interviews with a variety of stakeholders within the Lebanese electricity industry. This included representatives of public and private electricity companies, civic and industry associations, political parties, government and municipal officers, consultants, informal entrepreneurs operating the emerging network of generators and electricity users. These interviews were complemented with an analysis of official documentation, data provided by the utility company and a review of the Lebanese press and blogosphere relating to this issue since 2005. This analysis helped to capture the discourses justifying policy interventions, supported by mapping key controversies, and identified the most visible forms of protest.

Energy Inequalities, Street Politics and Urban Grid Politics

Infrastructure, being at the core of the production and sustenance of the city, is directly involved in the production of urban inequalities and thus in the shaping of social and political hierarchies (Swyngedouw, 2004; McFarlane & Rutherford, 2008; Graham, 2010). Without rejecting the structural dimensions of such domination, scholars have stressed that in everyday life such processes of establishing social and political hierarchies do not go unchallenged. This pervasiveness of contestation demands a reorientation of research towards the agency of 'minor' actors, the sites and moments where and when protest and dissent is enacted, and how the political order can be challenged, shaken and even reversed (at least partially and temporarily), as occurred during the Arab Spring (Mitchell, 2002; Parker, 2009; Bayat, 2012; Bayat, 2013). In the Arab world and beyond, street politics have emerged as a highly visible way of staging protest, with major squares and arteries becoming a symbol of the revolts (Swyngedouw, 2011). Yet, contesting the political order, and urban injustice through this, is not only a matter of street protest. It also involves dissenting through other forms of action (Bayat, 2009; Allegra et al., 2013): a more ordinary form of street politics (Bayat, 2012), often intersecting with a politics 'from within the energy system' (Mitchell, 2011: 12).

Timothy Mitchell's (2011) book *Carbon Democracy* offers a useful framework to unpack the links between the specific materiality of dominant energies (such as coal or oil) and the democratic order. Mitchell (2011: 252–3) achieves this through 'follow[ing] closely a set of connections that were engineered over the course of a century between carbon fuels and certain kinds of democratic and undemocratic politics'. In this way, 'exploring the properties of oil, the networks along which it flowed, and the connections established between the flows of energy, finance and other objects provides a way to understand how the relations between these various elements and forces were constructed'. In Mitchell's analysis of energy flows in the twentieth century, the concentrated nature of coal and the large workforce involved in its excavation and transportation gave workers the power to advance their democratic rights. Conversely, the greater fluidity and lightness of oil, together with the lower number of workers involved, allowed firms and oil states to circumvent and counter the power of democratic forces. Such mapping of energy circuits provides insights into the political and contested nature of the connections that allow the continued circulation of energy and capital at different points in the networks. A key advantage of Mitchell's analysis is the unravelling of the vulnerability or precariousness of these connections, to be understood as socially constructed and inherently political. Cupples' (2011) analysis of protests against the privatisation of an electricity distribution utility in Nicaragua intersects and helps to transpose Mitchell's work to the city scale. Like Mitchell, Cupples emphasises the political effects of the materiality of the network. Here this materiality plays out in various settlements as customers 'tactically and creatively enrol the nonhumans' by tampering with electric meters, so as to disrupt a device that both symbolically and materially enacts the commodification of electricity (Cupples, 2011: 945).

In transposing the framework advanced by Mitchell and Cupples onto the urban level in Amman and Tunis, recent research has emphasised the vulnerability of a specific segment in electricity circuits: the 'end of line' (Verdeil, 2014). Comprised of transformer stations, poles and meters, this segment of the electricity infrastructure enables various actors to deploy their agencies and interact with the grid. In the context of the material and social vulnerability of the 'end of line', utility workers repair circuits, collect bills and fight illegal connections whilst dwellers and customers hook up lines and tamper with meters. Workers also use their strategic position to disrupt the circuits of money and energy when they strike (Verdeil, 2014). This account of the 'end of line' is also applicable to the case of Beirut, yet it needs to be refined in two directions. First, the embeddedness of such struggles in urban space as well as of their political meaning and reach needs further consideration. While Cupples (2011) describes anti-neoliberal campaigning targeting electric meters, Bayat (2009) offers a wider understanding of street politics encompassing not only protests staged on streets and squares with explicitly political slogans but also ordinary practices in 'back alleyways' and popular neighbourhoods. These everyday practices, which he calls 'the quiet encroachment of the ordinary', may express more dissent than open protests; they are enacted through 'non-movements', or 'collective actions of non-collective actors' (Bayat, 2009: 14). In this respect, they are less politicised – which does not mean non-political. Following this line of thought, this chapter emphasises the forms of these political mobilisations (collective/group-specific/individual) and their democratic content, that is, their capacity to advance social rights and to define a common future.

Second, it is necessary to go beyond the very material end of the grid (the 'end of line'), and understand not only the politics of disrupting circuits but also that of imposing one's power on the grid. Mitchell (2011), analysing the geopolitics of oil production and supply, shows how energy firms built cartels in order to control supply and through this oil prices. Such form of control over the network could succeed only thanks to alliances with actors – external to the energy circuit – that exert forms of sovereign control over the territory crossed by the flow of oil. These (imperial and business) actors helped dismiss attempts at building an international and democratic sovereignty over the oilfields at the end of the First World War and during the Interwar period, continuing the control and expanding repression of workers' unions. These powers, such as the royal family of Saudi Arabia, are to this day considered to be among the least democratic forces in the world. These notions of alliances, of building local power over territories to enforce the political security of energy circuits, and the regressive politics this entails, is used to analyse the alternative electricity grids that Beirutis resort to during power cuts, since they are organised as small local-level monopolies.

In what follows, the chapter successively examines the uneven geography of accessing electricity and the associated establishment of relationships of domination linked to supply within the Lebanese electricity crisis. It also looks at strategies of resistance to domination based on a struggle around the control of the electricity grid and the new relations of power built around emerging informal territorialised electricity grids that spread across the city. This highlights the various ways in which urban electricity is politicised, and if and how it contributes to reconfigure the existing political order.

Uneven Geography of Electricity Supply and the Sectarian Order

For years, an electricity crisis has been at the core of Lebanese politics. In 2010 the electrical output of EDL, the national public utility, amounted to about 1,500 MW – significantly less than the estimated 2,300 MW required to satisfy demand (Bassil, 2010). The electricity supply suffers from heavy shortages and long lasting power cuts on a daily basis (8–12 hours a day on average in 2012). EDL records heavy annual losses ($1.5 billion in 2008–10), constituting a major part of the ballooning state deficit. On the one hand, the rising price of imported fuels is not reflected in the tariffs paid by customers. On the other, rampant corruption, theft and poor collection rates aggravate the company's financial burden (Verdeil, 2009; Hasbani, 2011). Numerous policies have attempted to reform the sector and since 2002 successive governments have discussed the privatisation of the electricity sector. However, until 2011 the key interventions in the state-owned utility were limited to reducing investments and retrenching the workforce as the company moved to using contractual workers, often employed through clientelistic channels.

The current electricity supply in Lebanon is marked by strong inequalities, particularly noticeable through both an uneven geography of access and the unequal subsidies that several groups of customers enjoy. Such an energy configuration further reinforces the country's existing inequalities, while solidifying the sectarian divide that characterises Lebanese politics. The most striking feature of this uneven supply lies in the uneven service which favours Beirut over other parts of the country. Since 2006, Beirut receives about 21 hours of power a day, while elsewhere in the country daily power cuts can reach 12 to 16 hours. The contrast is particularly felt in the city's immediate suburbs, differentially affecting people living in the same urban fabric since areas outside the municipal boundary – from one side of a street to the other – live with very different conditions of electricity access and supply (Figure 8.2).

Favouring Beirut over the rest of the country is undoubtedly a political choice, though no reason has ever been publicly given. From the point of view of the utility, this choice secures its financial income since Beirut's customers are responsible for the highest share of electricity consumption in the country, with little electricity theft and non-payment recorded there. The most obvious economic implication of this arrangement is the protection of the country's main economic sectors (banks, administration and hotels) from power cuts, in this way securing the private interests of their owners. Yet the choice of prioritizing Beirut's municipal boundary can also have political undertones. Between 2009 and 2013 the Ministry of Energy was in the hands of the Free Patriotic Movement (FPM), a Christian party claiming to rationalise the management of the sector. Allied with the Shia parties of Amal and Hezbollah, the FPM claimed – and seemed to have – the political means to push for a more balanced distribution of supply. This move would have also fitted the political interest of the alliance, since all three parties had their power bases in the suburbs (north and east for the FPM; south for the Shia). But the national level Council of Ministers rejected two such attempts. This failure highlights the complex nexus of business and elite interests inside the governing coalition that resisted the change.

Technical factors also play a role in the configuration of an uneven geography of electricity, aggravating the already unbalanced official supply schedule. Indeed, lack of investment in infrastructure alongside sprawling and overcrowded

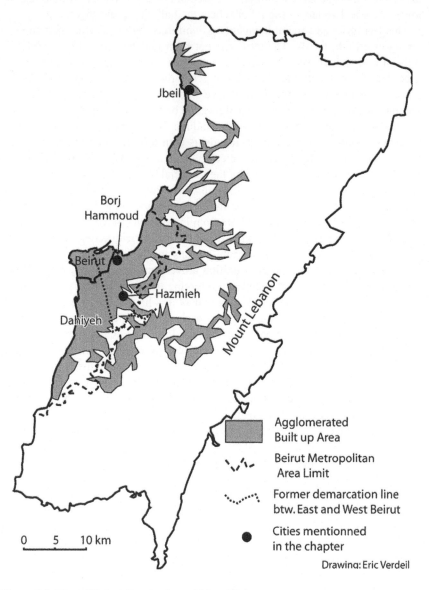

Figure 8.2 Map of Beirut, its suburbs and Mount Lebanon
Source: the author

suburban sectors during and after the civil war have resulted in an undersized distribution grid, increasing local shortages. For example, in Dahiyeh, the southern suburb of Beirut, technical limitations restrict supply whilst the not so distant city centre enjoys extensive surplus capacity (Wehbe, 2012). As a consequence, Beirut dwellers – on average the wealthiest in the country – are also the best supplied; the lower-income dwellers of the suburbs bear the bulk of the shortages.

This territorial and social inequality is reinforced by the subsidy system governing electricity tariffs. The tariff has not been updated since 1996, while the price of fuel – purchased on the international market – has more than quadrupled. Electricity is sold well under the cost of generating it (Bassil, 2010). This means that those who consume the most receive a correspondingly larger subsidy from the state budget. Beirut dwellers, wealthier and better supplied than their suburban counterparts, enjoy a larger share of this indirect form of government aid, though no data is available to estimate it. Residents with infrequent and disrupted electricity are increasingly forced to use costly generators (World Bank, 2009). The resulting imbalance is augmented by the financial burden that those with limited and intermittent electricity access have to bear in order to alleviate power cuts: purchasing electricity from privately owned generators running on diesel, the cost of which is non-subsidized and suffers from international price fluctuations.

In Beirut, the uneven supply and distribution of electricity across the city region has prompted numerous popular mobilisations in public spaces. Yet such protests have never coalesced into a unified movement; instead, they have served sectarian and local political agendas, reinforcing the city's current political fragmentation. Street demonstrations against darkness have taken two forms. First, and less frequent, are big demonstrations backed by political parties at symbolic urban spaces. This form of protest can be exemplified by a demonstration that took place in Chiyah, on the former demarcation line between East Christian Beirut and the Southern suburbs, on the 27th of January 2008. This was mostly attended by Shia people from Dahiyeh, probably those most affected by the power cuts at the time. The protest resulted in riots and other forms of violence, with nine deaths and accusations between opposing political parties about the cause of the conflict. Government backers claimed that Hezbollah, the main political force in Dahiyeh acting at a time of fierce political struggle, wanted to use this event as a way to pressure the government, in effect overlooking the reality of the uneven service supply and the legitimacy of the protesters' claims. Yet Hezbollah itself, apparently losing control over the mobilisation, made efforts to calm down popular anger and reaffirm a patron-client relationship by bringing in new generators to the suburb (Chit, 2009).

The second form of protest against darkness relates to small neighbourhood-scale protests where mobs, apparently without the formal backing of political parties or other organisations, disrupt mobility and circulation in the city by burning tyres in important thoroughfares. In their demands, protesters, mostly young and deprived, make few connections with wider economic and social issues such as joblessness. The political elite respond to such protests in sectarian and divisive terms. For example, the response of the Minister of Energy and Water, Gebran Bassil, to the street mobilisations that characterised the heat wave of August 2010 reveals a common type of sectarian politics when dealing with the uneven condition of service access in the city. Bassil, a member of the Christian FPM, declared: 'In some districts, protests against power cuts are politically motivated'; slamming

Muslim districts, he added, 'it is not permitted for a region where illegal hook-ups multiply to protest against power cuts. . . . It appears that Christian regions pay their due more than others, but it's the truth' (L'Orient-Le Jour, 2010).

These sectarian narratives, which permeate public discussions about the city's unequal access to basic services, overlook a more nuanced account of the geography and sociology of electricity fraud; one that highlights the role of local political leadership in covering illegal grid connections (Verdeil, 2009), and one which, stepping away from blanket approaches, considers the quantities consumed. In 2009 the Head of the Higher Privatisation Committee, a government body established to support the privatisation of a variety of public services and entities, acknowledged that 'richer customers that have higher electricity bills cost the state more money' than illegal dwellers who 'steal to light a small apartment' (Hayek, 2009: 31). Several industrial operations, leisure clubs and even political leaders have been singled out by the press for illegal electricity hook-ups or lack of bill payment. This acknowledgement by the Higher Privatisation Committee also means that fraudulent big customers, even in small numbers, are responsible for a large amount of lost income for the electricity utility, outweighing 'small fry' fraudulent customers whatever religion they belong to. However, like with many other social struggles in the country (Abi Yaghi & Catusse, 2011), the sectarian discourse prevents an alternative approach to electricity loss and access, particularly obscuring the class dimensions of the problem.

Grid-Embedded Resistance

The lack of efficacy of voicing dissent in the public sphere prompts an examination of other forms of politicisation of electric issues that are emerging in the current crisis: the use of the material and social configuration of the grid itself in order to advance social and political claims around service provision. In these situations, actors involved in conflicts with the state utility try to reap advantage from the technical-political vulnerability of urban electricity circuits. In contrast with the more universal claims for access and equality examined in the previous section, here claims tend to be more group or place-specific. Not all of these forms of resistance are successful, and the struggles appear to fail to open new democratic horizons for all citizens. Conversely, such forms of grid protest can feed more divisive politics, be it sectarian or class based.

One example of such grid-embedded resistance is provided by an examination of how workers in the electricity distribution sector exert a form of power that originates from their strategic location and function inside the grid. Between May and September 2012, EDL contractual workers – enrolled on a daily basis and without social benefits – fought for more than four months against a law that would contract electric distribution to private firms and turn them into employees of these new businesses. Workers claimed a right to be public employees officially working for the publicly owned utility. They feared the law was a first step towards the dismantling and privatisation of the utility company. In response, they implemented a strike during the summer months, a time where the usual power cuts are particularly long and heavily felt. Workers stopped repairing disrupted infrastructures damaged because of strong electricity demand, further aggravating the power cuts experienced by users. They retained payments collected from customers, causing the company an additional loss of income and preventing it from buying the fuel required to generate electricity. In their actions, they were making use of their position as indispensable pivots in the operation of the grid. Though their fight was unpopular and they received no support from the electricity union (which limits its representation to workers that enjoy the status of civil servants), their actions enabled them to reach an agreement with the government. Contractual workers were promised a facilitated recruitment procedure for their formal integration within the utility. Whilst this struggle had a clear class dimension, its sectarian dimension cannot be dismissed: since the majority of the workers belonged to the Shia denomination of Islam, Christian politicians opposed the agreement with the argument that it would create a sectarian imbalance.

In contrast to that workers' struggle, the everyday fights of dwellers in some informal neighbourhoods to secure and defend electricity access proves to be much more precarious and politically disappointing: the tampering of electricity meters. The 'end of the grid', where electric power is transformed from medium into low voltage, is a liminal space where connections between utility networks, customers and electricity workers take place. Its sub-stations and transformers are physically located within the neighbourhoods, and whilst entering (or tampering with) them can be dangerous, they are easier to access than any other segments of the grid. After the transformers, electric power flows through low-voltage lines towards the dwelling, and skilled people can relatively easily hook up illegal or informal connections (Zaki, 2011). The connection materialises once again through the electricity meter, which itself can be targeted, bypassed and tampered with in various ways. Tampering with meters a common practice in Lebanon since the civil war. Non-technical losses, once estimated over 50 per cent (Badelt & Yehia, 2000), now amount to about 23 per cent of the total output (Hasbani, 2011). However, though the utility has taken significant steps in order to eliminate meter tampering, the practice increases each time the political bickering in the country results in a governmental stalemate. They are actions that take advantage of the physical vulnerability of the network, despite new anti-theft technologies such as twisted wires and sealed meters which, after the civil war, made illegal hook-ups more difficult.

Yet, the connection between the network and the customer is not configured solely through physical devices. It also materialises in the periodic meeting of collector and customer, when the collector reads the meter and distributes the bill to the customer. Here again the customer can display its power by refusing to pay in time. In doing so, they expose the household to the possibility of being cut off – excluded from the grid. But such a move can only be achieved through a team of utility workers capable of visiting the building for the purpose of physically disconnecting the line. Since broader power dynamics frame these relationships, these micro-resistances and the possibility of disconnection cannot be understood outside of the overlapping political geography of the city. Such practices of resistance, for instance, are easier in areas that enjoy political backing by local forces usually connected to powerful parties. It means that grid-embedded resistance is entangled in classical patron-client relationships. Sometimes, even dwellers that don't enjoy such protection oppose a physical resistance to the anti-fraud teams. The press regularly reports on such incidents (L'Orient-Le Jour, 2004; The Daily Star, 2009). In such cases, Bayat's suggestion of 'politics of redress', or the collective action required in order to defend the gains achieved through the 'quiet encroachment of the ordinary' (Bayat, 2009: 58), may be useful. In troubled political times, as has been the case in Lebanon since 2005, these resistances have prevented or limited crackdown campaigns, a phenomenon acknowledged by state authorities. For instance, on the 19 December 2006, *L'Orient-Le Jour* cited Mohamad Safadi, acting Minister of Energy, speaking of an 'increase of power theft from the grid, [alongside] EDL employees being unable to report or monitor the infractions, given the current situation in the country' (L'Orient-Le Jour, 2006).

In contrast to the Nicaraguan campaign of resistance against privatisation described by Cupples (2011), tampering with electricity meters in Beirut is a form of grid resistance that is not explicitly accompanied by clear demands expressing a political meaning. Dwellers in Raml al-'Ali, one of the poorest informal neighbourhoods in the southern suburb of Beirut, reported poverty and the high cost of electricity as the main reason for illegal (or pirate) electricity hook-ups (Khayat, 2008). Dwellers in the neighbourhood did not expect an improvement in the provision of electricity, and made do as they could. Khayat's survey highlighted a lack of solidarity between dwellers regarding hook-up practices. For instance, they prefer to hook-up to a wire already illegally connected to the grid in order not be ticketed. Dwellers had frequent disputes about the responsibility of incidents caused by hook-ups – such as overloads, which can create long lasting power cuts – because the utility company did not provide its repair services in such cases in order to 'punish the pirates', as an engineer once expressed (interview, Beirut, June 2006).

The capacity of electricity workers to disrupt the system and gain bargaining power to improve labour conditions, coupled with the vulnerability of the modern metropolis to power cuts, echoes Timothy Mitchell's (2011) arguments around the power of coal workers and their ability to fight for democratic rights. Similarly, the dwellers' strategy of 'pirate' hook-ups and meter tampering rests on acquiring a power of action 'from within the new energy system' (Mitchell, 2011: 12). Both agents target vulnerable points in the electricity circuit of the city, and exploit a favourable balance of power in certain places. Yet these forms of resistance are precarious and often depend upon patron-client relations that forgo the possibility of structural improvements. Moreover, it appears that such actions do not express collective strategies and rely mostly on individual and opportunistic behaviour, rather than on a collectively devised project of enhancing rights to resource access.

The Geographically Splintered Politics of Informal Electricity Grids

This last section turns to another case of electricity politics in Beirut: the informal grids established around privately run diesel generators. This pervasive form of supply, which has spread since the civil war (Awada, 1988; Davie, 1991), now constitutes a consolidated form of service provision in Lebanese cities, foremost in Beirut and its suburbs. Born out of the need to rely on emergency sources during long power cuts, these electricity networks proliferated initially as individual solutions and later as collective responses operating at the scale of the building of private firms and even entire neighbourhoods. Originally run in a spirit of solidarity, they have gradually taken a commercial orientation with segregating effects. By 2007 and 2008 it was estimated that private generation in Lebanon amounted to 34 per cent of total electricity production, with 56 per cent of households relying on private suppliers for lighting (Central Administration of Statistics, 2008; World Bank, 2009). Since interruptions in the average daily supply of electricity have only worsened, particularly in Beirut where people used to enjoy almost full supply until 2006, this figure has probably increased. The use of electricity generators and their associated micro-grids provide a relevant case study to evaluate the governance of the new social relations shaping resource vulnerability in the city and the failures of its infrastructural network.

Widely used, private electricity generators and their grids are not only temporary and superficial fixes to the daily interruptions of service characteristic of the city. They have constituted an infrastructural geography deeply embedded in the everyday, which restructures the urban fabric according to its material, social and political characteristics and produces its own form of politics. These wire networks servicing dwellings are spread chaotically from generators located in convenient places in the neighbourhood – such as vacant spaces or abandoned factories. At the dwelling, they connect to the internal network via a manually operated switcher that is activated whenever there is a power cut (Figure 8.3). Gradually, these informal grids have concealed new devices, slowly embedding them into other urban objects. For example, in new buildings, a secondary network parallel to the official one runs inside the walls, using automatic switchers that enable connection and disconnection from the formal grid as required. Marketing material for new gated communities in Beirut speak of 24/7 water and electricity delivery as distinctive features, achieved thanks to generators run by real estate developers. The generator has been normalised as a basic service, invisible to the user, just like public infrastructure has been black-boxed and buried over time (Kaika & Swyngedouw, 2000). In new buildings, notably in high-end neighbourhoods, the arrangement of generators is embedded within the overall building design, in this way avoiding or limiting nuisances such as the smoke of the generators for final users (Figure 8.4). This trend toward business-oriented grids has spurred the development of new commercial services, for instance, through differentiated levels of supply – exhibiting an ability to incorporate and respond to the variable needs and capacities of, for example, smaller or poorer customers. Consistent with the advancement of a splintering urbanism (Graham &

Figure 8.3 The grid of a generator in Zaatriyeh, a popular suburb east of Beirut
Source: the author

Marvin, 2001), the new grids mirror existing social hierarchies and, to an extent, reinforce the city's inequalities.

The emergence of these new infrastructural capabilities has produced its own politics. The state, still promising 24/7 electricity, has never recognised the

Figure 8.4 A generator serving a middle class building in Beirut, 2014
Source: the author

generators sector; yet, it has never repressed it. Generators, as 'grey spaces', remain between the whiteness of the law and the darkness of what is supposed to be demolished or expelled (Yiftachel, 2009; Gabillet, 2010). Such grey spaces are temporary, at the same time tolerated and condemned, always waiting to be cleared. As such, they can be used as a tool for governing without granting rights to users; there are no rights attached to a service that should not exist. Electricity via these small generators has become a profitable business abiding to no rules. Depicting generator businesses as mafias is common (Batah, 2011; Mohsen, 2012), a reference related both to the nature of their profits – which escape any kind of tax and state regulation – as much as to the abuses committed against clients and competitors. The businesses tend to organise spatially as a local monopoly, are defended by violent means and clients have little recourse against wrongdoings such as unjustified pricing and defaults. Rumours of politicians being involved in the business in their regional strongholds are numerous. Turning back to Mitchell (2011), who emphasises how building monopolies operate as a means to control output and price on the energy market, it is clear that in Beirut this cannot be done solely from 'within the network' but only through alliances with local political forces and the exploitation of local power configurations.

The case of Hazmieh, a middle class suburb of about 30,000 inhabitants in the southeast of the capital, illustrates a common situation where the owners of generators operate their business with the backing of the municipality under a laissez faire approach open to bribes and corruption. Prices for electricity are higher here than in other parts of the city. Voter power and the possibility of pressure from the electorate for the regulation of generators is out of the question, since most citizens only recently settled in the suburb and are not registered on the local voting lists (Favier, 2001; Verdeil, 2005). This leaves residents vulnerable to higher electricity costs, a common condition in Lebanon. The case of Jbeil illustrates another example of these electric politics (Gabillet, 2010). This town of about 40,000 inhabitants is less a suburb of Beirut than a small tourist city in the outskirts of the metropolis. There, the firm *Karhaba Jbeil* (Electricity of Jbeil) holds a private concession dating back to the French mandate, formally replacing EDL as supplier of electricity. Several informal generator businesses operated in the city, but a mayor elected in 2004 decided to clear the landscape of these technologies in order to beautify the city. He pressured the operators of generators to withdraw their electricity from the market by blocking their use of the poles that belonged to the municipality. Instead he pushed *Karhaba Jbeil* to become a monopolistic actor. The mayor was able to deliver this policy through a deal with the supplier to keep the electricity costs low and forestall attempts to re-install the generators.

The case of Borj Hammoud, an Armenian suburb just outside Beirut with about 100,000 inhabitants, offers another illustration of the strong role municipalities can play in regulating the private grids (Gabillet, 2010). Since 2010, in response to calls by civic and consumer associations for regulating the sector, the Ministry of Energy issues a monthly suggested tariff for their output and leaves open the option for municipal governments to enforce this pricing. Borj Hammoud has been among the pioneers. Local police have been enrolled to monitor the generators' output. Both the use of public poles and the location of generators in public spaces are negotiated between owners and the municipality. But the negotiation also includes electricity prices. Decisions are made public through a local Armenian TV channel. This innovative municipal involvement in the emerging micro-grid of the generators rests on re-taking control of the grid, via poles, wires and even meters for each generator. Arguably, this type of municipal involvement is also linked to the political agenda of the Tachnag Party, the major Armenian political power in the country, to maintain a local political hegemony and thus to prevent conflicts around an important issue for its constituents. Such processes are less about granting rights than de facto ruling a service that remains outside of the remit of the law.

The actions of Borj Hammoud lay at the intersection of holding power on the grid and a territorial power that not only permits the enforcement of the monopoly but also regulates relations between producers and customers. The case of Hazmieh is one where the owners of the informal grid operate with the support of a local power, whereas the case of Jbeil illustrates a more interventionist approach, but where the private firm is given free rein. In all cases, citizens have limited rights and few means to defend themselves against the power of these new grids. Local voting rights provide little leverage, given the disconnection between the place of residence and the place of voting, a common issue in Lebanon. One of the only situations where citizens can exert their rights is in the context of a commercial relationship, as is the case of high-end condominiums and gated communities. Here neighbourhood associations can, for instance, cancel the services of a service provider whose performance does not fit the agreed standard. But such actions are restricted to a wealthy minority (Glasze, 2003). The politics created by private energy grids are not only highly uneven between social classes, but also geographically fragmented according to specific local political contexts. This highlights the impossibility of a shared governance of electricity at the level of the city, illustrating, to return to the cartoon of Abdelhalim Hammoud (Figure 8.1), the infrastructural disintegration and meltdown of the country.

Conclusion

This chapter mapped the disruptions and reconfigurations of electricity circuits in the Greater Beirut region, and demonstrated how such disruptions both reflect and create particular configurations of power. To do so, it built on Timothy Mitchell's (2011) project to follow the tracks of energy, their vulnerabilities and the struggles that take advantage of them. It also built on Cupples' (2011) commitment to unravel the precarious agencies involved and the enrolment of material devices in the process of advancing social rights. The analysis allowed an understanding of the kind of politics such endeavours produce: the manifestation of discomforts via protests and dissent in the public space, highlighting collective, 'non-collective' (Bayat, 2009), corporatist and individual forms.

Through this conceptual approach, the chapter discussed what such mobilisations managed to achieve, alongside their political reach in view of advancing democratic rights. This can be summed up through a typology of grid politics based on their ability to disrupt and alter the power relations built in and around electricity supply (Table 8.1). A central concern here is to reflect on the specificity and efficiency of forms of political action which, paraphrasing Mitchell and

Table 8.1 The various types of grid politics in Lebanon

Cases examined	Type of actors	Type of claims	Form of politics (street/ from within the grid)	Political horizon of the struggle
Burning Tyres/Mass demonstration	Citizens/ customers	Right to light	Street protest	Sectarian politics
Utility work- ers' movement	Workers	Right to formal job (job tenure + benefits)	Disrupting the grid (+ street protest)	Defence of the national utility
Informal dwellers' hook- ups and meter tampering	Customers	Access to light (not claimed as a right)	Individual quiet encroach- ment/collective politics of redress using power from within and around the grid	Overcoming and defend- ing local advantages
Regulating neighbourhood generators	Municipality	Access to light (not granted as a right)	Control on the grid	Local poli- tics (often sectarian)

Cupples, work from within the energy system and tactically and creatively enrol the nonhumans. These have varying degrees of success in advancing democratic rights in the face of expanding forms of privatisation and, more globally, the sub-jugation of citizens to a prevailing unequal socio-political order.

In the context of Beirut, the typology distinguishes four key domains. The first one is the politicisation of electricity provision through various kinds of street protests. This fails to exert sufficient pressure on the political order and is handled rather through sectarian means. The second domain, the struggle of workers, high-lights the power of those who, operating from within the power grid, disrupt flows of electricity and capital, and therefore exploit to their advantage the fact that electricity is an indispensable infrastructure of life in the metropolis. Just as in the case analysed by Mitchell (2011), such conflicts can advance democratic rights. In parallel, these workers claimed the defence of a national cause – the public role of a national utility, threatened by unbundling and privatisation in the electricity sector.

A third domain of politics, the struggle to access electricity through illegal means in remote suburbs, illustrates an attempt to use the same kind of power from 'within the energy system'. But in contrast with the previous case, these achieve-ments are more precarious. They depend upon local political support – often involving sectarian politics – or can be sustained only as long as the political stale-mate lasts. It is more an opportunistic way for disenfranchised people to access free electricity in a particular local context rather than a struggle with a broader social scope. Such fights are plagued by internal strife and a lack of solidarity.

A fourth and final domain of politics analysed in the chapter is related to the informal grids supplying power from generators. These grids echo other informal infrastructure services in cities in the global South, with two distinct features. First, they do not generate a place-specific form of service provision, to be found only in poor areas bypassed by infrastructure. Since power cuts are more or less ubiquitous (though their duration reflects uneven access), private grids are not located in specific areas but everywhere in the city. Second, they coexist with the main electricity grid, using some of its infrastructure, notably poles and neutral lines. Therefore, they are not class-specific, as both poor and rich makes use of such informal infrastructures. One of the more salient political features of such a networked system lies in the emergence of a violent localised form of electric capitalism, constructing and exploiting the position of power that local monopolies grant to their private owners. Only the municipalities can challenge this, thanks to their level of territorial control and, hence, control of the grid itself. This case highlights the interface between territorial power and the power from within the grid. In some configurations, local policies can be less detrimental to the citizens-customers, though it cannot be conceived of as really democratic but rather less regressive. In the Lebanese case, nevertheless, the splintering dynamics that are at work at the local scale do not produce any converging direction in these struggles and are blurring any shared metropolitan and national horizon.

The political struggles examined in this chapter highlight the deep inequalities produced by the long lasting electricity crisis in Beirut. Achieving energy justice, particularly where democratic forms of politics are significantly impeded, can pass through other kinds of politics, combining grid power and local territorial power in order to disrupt circuits, redress eviction threats or protect advantageous configurations of access. This involves scrutinizing the minor human and non-human agencies that are too often overlooked or dismissed. Nevertheless, these forces remain entangled in divisive schemes of power that prevent them from achieving any major reshuffling of the political system or long lasting changes to the electricity system.

Acknowledgements

The author would like to thank Mona Fawaz, Andrés Luque-Ayala and Jonathan Silver for generous comments, suggestions and editing at various stages of the writing of this chapter, as well as Alain Nadai for his insistence on thinking about the politics of the generators.

References

Abi Yaghi, M.N. and M. Catusse, 2011. 'Non à l'Etat holding, oui à l'Etat providence!'. *Revue Tiers Monde* 206: 67–93.

Allegra, M., I. Bono, and J. Rokem, 2013. 'Rethinking Cities in Contentious Times: The Mobilisation of Urban Dissent in the "Arab Spring" '. *Urban Studies* 50(9): 1675–88.

Awada, F., 1988. *La gestion des services urbains à Beyrouth pendant la guerre: 1975–1985.* Talence: CEGET.

Badelt, G. and M. Yehia, 2000. 'The Way to Restructure the Lebanese Electric Power Sector: A Challenge for the Transitional Management'. *Energy Policy* 28(1): 39–47.

Bassil, G., 2010. *Plan stratégique national pour le secteur de l'électricité*. Beirut: Ministry of Water and Energy.

Batah, H., 2011. Policy Powers Gray Market in Generators. *The Daily Star*. [Online]. 12 September. Available from: http://www.dailystar.com.lb/Business/Lebanon/2011/Sep-12/148493-policy-powers-gray-market-in-generators.ashx#axzz1Xoo6s0Bz. [Accessed: 7 May 2015].

Bayat, A., 2009. *Life as Politics: How Ordinary People Change the Middle East*. Stanford: Stanford University Press.

Bayat, A., 2012. 'Politics in the City-Inside-Out'. *City & Society* 24(2): 110–28.

Bayat, A., 2013. 'The Arab Spring and its Surprises'. *Development and Change* 44(3): 587–601.

Bickerstaff, K., G. Walker and H. Bulkeley (eds.), 2013. *Energy Justice in a Changing Climate*. London: Zed Books Ltd.

Central Administration of Statistics, 2008. *Living Conditions Survey 2007*. Beirut: Central Administration of Statistics.

Chit, B., 2009. Divisions confessionnelles et lutte de classe au Liban. *Que Faire ?* n°10, January–March. [Online]. Available from: http://quefaire.lautre.net/que-faire/que-faire-lcr-no10-janvier-mars/article/divisions-confessionnelles-et. [Accessed: 7 May 2015].

Coutard, O., 2008. 'Placing Splintering Urbanism: Introduction'. *Geoforum* 39(6): 1815–20.

Cupples, J., 2011. 'Shifting Networks of Power in Nicaragua: Relational Materialisms in the Consumption of Privatized Electricity'. *Annals of the Association of American Geographers* 101(4): 939–48.

Davie, M.F., 1991. La gestion des services urbaines en temps de guerre, circuits parallèles à Beyrouth. In: N. Beyhum (ed.), *Reconstruire Beyrouth: les paris sur le possible: table ronde tenue à Lyon du 27 au 29 novembre 1990*. Lyon: Maison de l'Orient, pp. 157–93.

Favier, A. (ed.), 2001. *Municipalités et pouvoirs locaux au Liban*. Beyrouth: Centre d'études et de recherches sur le Moyen-Orient contemporain.

Gabillet, P., 2010. 'Le commerce des abonnements aux générateurs électriques au Liban'. *Géocarrefour* 85(2): 153–63.

Glasze, G., 2003. 'Segmented Governance Patterns – Fragmented Urbanism: The Development of Guarded Housing Estates in Lebanon'. *The Arab World Geographer* 6(2): 79–100.

Graham, S. (ed.), 2010. *Disrupted Cities: When Infrastructure Fails*. London: Routledge.

Graham, S. and S. Marvin, 2001. *Splintering Urbanism: Networked Infrastructures, Technological Mobilities and the Urban Condition*. London: Routledge.

Hasbani, K.U., 2011. Electricity Sector Reform in Lebanon: Political Consensus in Waiting. *CDDRL Working Paper 124*. [Online]. Available from: http://iis-db.stanford.edu/pubs/23465/No._124_Electricity_Sector_Reform.pdf. [Accessed: 7 May 2015].

Hayek, Z., 2009. 'The Monthly Interviews the Secretary General of the Higher Council for Privatization, Ziad Hayek'. *The Monthly* (87): 30–1.

Heynen, N.C., M. Kaika and E. Swyngedouw, 2006. Urban Political Ecology. Politicizing the Production of Urban Natures. In: N. Heynen, M. Kaika and E. Swyngedouw (eds.), *In the Nature of Cities*. London: Routledge, pp. 1–20.

Jaglin, S. and É. Verdeil, 2013. 'Énergie et villes des pays émergents: des transitions en question. Introduction'. *Flux* 93–94(3): 7–18.

Kaika, M. and E. Swyngedouw, 2000. 'Fetishizing the Modern City: The Phantasmagoria of Urban Technological Networks'. *International Journal of Urban and Regional Research* 24(1): 120–38.

Khayat, N., 2008. *Self-help Urban Services in Informal Settlements. The Case of Water and Electricity Services in Raml al-'Ali in Beirut*. Master Thesis. American University of Beirut.

L'Orient-Le Jour, 2004. Social – L'armée et les FSI accueillies à coups de pierres et de bâtons. Des émeutiers de Jnah empêchent l'EDL de faire son travail. *L'Orient-Le Jour*. [Online]. March 9, http://www.lorientlejour.com/article/465905/Social_-_L%27armee_et_les_FSI_accueillies_a_coups_de_pierres_et_de_batonsDes_emeutiers_de_Jnah_empechent_l%27EDL_de_faire_son_travail.html. [Accessed: 9 March 2004].

L'Orient-Le Jour, 2006. EDL le vol de courant est en hausse, souligne Safadi. *L'Orient-Le Jour*. [Online]. 19 December. Available from: http://www.lorientlejour.com/article/547742/EDLLe_vol_de_courant_est_en_hausse%2C_souligne_Safadi.html. [Accessed: 19 December 2006].

L'Orient-Le Jour, 2010. Le ras-le-bol de Gebran Bassil: Ceux qui protestent contre les coupures sont ceux qui volent le courant'. *L'Orient-Le Jour*. [Online]. 20 April. Available at: http://www.lorientlejour.com/article/668544/Le_ras-le-bol_de_Gebran_Bassil+%3A_Ceux_qui_protestent_contre_les_coupures_sont_ceux_qui_volent_le_courant. [Accessed: 20 August 2010].

McFarlane, C., 2010. Infrastructure, Interruption, and Inequality: Urban Life in the Global South. In: S. Graham (ed.), *Disrupted Cities: When Infrastructure Fails*. London: Routledge, pp. 131–44.

McFarlane, C. and J. Rutherford (eds.), 2008. 'Political Infrastructures: Governing and Experiencing the Fabric of the City'. *International Journal of Urban and Regional Research* 32(2): 363–451.

Mitchell, T., 2002. *Rule of Experts: Egypt, Techno-politics, Modernity*. Berkeley: University of California Press.

Mitchell, T., 2011. *Carbon Democracy: Political Power in the Age of Oil*. New York: Verso.

Mohsen, A., 2012. Lebanon's Electricity Mafia. *Al Akhbar English*. [Online]. 29 June. Available from: http://english.al-akhbar.com/content/lebanon%E2%80%99s-electricity-mafia. [Accessed: 29 June 2012]

Monstadt, J., 2009. 'Conceptualizing the Political Ecology of Urban Infrastructures: Insights from Technology and Urban Studies'. *Environment and Planning A* 41(8): 1924–42.

Parker, C., 2009. 'Tunnel-bypasses and Minarets of Capitalism: Amman as Neoliberal Assemblage'. *Political Geography* 28(2): 110–20.

Rutherford, J. and O. Coutard, 2014. 'Urban Energy Transitions: Places, Processes and Politics of Socio-technical Change'. *Urban Studies* 51(7): 1353–77.

Swyngedouw, E., 2004. *Social Power and the Urbanization of Water: Flows of Power*. Oxford: Oxford University Press.

Swyngedouw, E., 2011. Every Revolution Has Its Square. In: M. Gandy (ed.), *Urban Constellations*. Berlin: Jovis, pp. 22–6.

The Daily Star, 2009. Attack on EDL Worker Causes Choueifat Power Cut. *The Daily Star*. [Online]. 22 August. Available from: http://www.dailystar.com.lb/article.asp?edition_id=1&categ_id=1&article_id=105591. [Accessed: 22 August 2009].

Verdeil, É., 2005. 'Les territoires du vote au Liban'. *M@ppemonde* 78(2): 1–24. Available from: http://mappemonde.mgm.fr/num6/articles/art05209.html. [Accessed: 22 January 2015].

Verdeil, É., 2009. 'Electricité et territoires: un regard sur la crise libanaise'. *Revue Tiers Monde* 198: 421–38.

Verdeil, E., 2014. 'The Energy of Revolts in Arab Cities: The Case of Jordan and Tunisia'. *Built Environment* 40(1): 128–39.

Wehbe, M., 2012. Solidere Burns Bright While Lebanon Goes Dark. *Al Akhbar English.* [Online]. 14 November 2012. Available from: http://english.al-akhbar.com/content/ solidere-burns-bright-while-lebanon-goes-dark. [Accessed: 7 May 2015].

World Bank, 2009. *Lebanon Social Impact Analysis – Electricity and Water Sectors.* [Online]. Social and Economic Development Group Middle East and North Africa Region. Available at: http://www-wds.worldbank.org/external/default/WDSContent-Server/WDSP/IB/2009/07/21/000333038_20090721001206/Rendered/PDF/489930ES W0P0891C0Disclosed071171091.pdf. [Accessed: 7 May 2015].

Yiftachel, O., 2009. 'Theoretical Notes On "Gray Cities": The Coming of Urban Apartheid?'. *Planning Theory* 8(1): 88–100.

Zaki, L., 2011. 'L'électrification temporaire des bidonvilles casablancais: aspects et limites d'une transformation "par le bas" de l'action publique. Le cas des Carrières centrales'. *Politique Africaine* (120): 45–66.

9 Barcelona

Municipal Engineers and the Solar Guerrillas

Anne Maassen

Introduction

The urban photovoltaic (PV) power station, towering over the mouth of the River Besòs in Barcelona, presents an arresting sight. While located on the fringes of central Barcelona, the Forum 2004 area's slogan is bearer of Barcelona's municipal authority's message of change: 'the future is now' (Agència d'Energia de Barcelona, 2003). It offsets about 440 tons of CO_2 every year and has the capacity to power 1,000 homes. Indeed, the 'pergola' is one of Europe's most iconic urban PV installations of the recent past, its installation coinciding with the Universal Forum of Cultures, an international cultural event which took place in Barcelona in 2004. Yet, while positioned by Barcelona's municipal authority as part of the city's new ecologically friendly urban 'energy culture', there are others for whom 'the sculpture is a monster' (Interview, 2009). For this activist and many others, the installation was little more than an orchestration of sustainability. The installation was seen as embodying the Council's orientation towards 'the outside', as one of many ways in which the Council has marketed Barcelona as a particular brand of sustainable city. A true icon of sustainability would look rather different for this group of people, who in this chapter are referred to as the 'solar guerrillas'.

This chapter is concerned with exploring the politics embedded in the shifting relationships around urban energy that took shape in Barcelona (Spain) in the first decade of the new millennium (approximately 1998–2010). During this period, Barcelona came to be celebrated as a 'pioneer municipality supporting energy sustainability' (PV Upscale, 2005), receiving most notably the European Commission's Local Energy Action Award in 2007 for the city's commitment to sustainable energy solutions. The award recognized a range of innovative urban planning policies and actions carried out by the municipality, the *Ajuntament de Barcelona* (Ajuntament, henceforth). These include a technically sophisticated urban energy strategy and a range of interventions principally involving different kinds of solar (thermal and electric) energy technologies. In parallel to the Ajuntament's award-winning policies, a set of Barcelona-based activists was undertaking a campaign with rather different ambitions around urban solar energy. Through a range of activities, they sought to expose and undo the systemic bias against renewable and decentralized power generation in Spain and involve regular citizens in energy generation; the ultimate aim being that of unsettling the unequal 'client' relationship between citizen and corporate interests. The activities of the solar guerrillas and those of the Ajuntament, respectively, can be seen as defining the dynamics in Barcelona's urban energy system in recent years. This chapter provides a window into the spatial politics of challenging the status quo of centralized fossil fuel and nuclear-based energy. In so doing, it first situates events in the context of urban technology scholarship and subsequently draws out how solar energy in Barcelona came to matter in more than symbolic terms – including its pivotal role in foregrounding and confronting the protagonists' divergent ambitions for change.

Obduracy and the Politics of Urban Technology

Substantial parts of the urban technology literature are concerned with the quint-essentially 'urban' technologies: the 'boring', mundane and largely 'invisible' technologies that 'provide water and energy, remove sewage and trash, deliver information, and transport us between homes and workplaces' (Hård & Misa, 2008: 8; see also Star, 1999). Features such as telephones, cars, residential, corporate and leisure facilities, buildings, sidewalks, public art and open spaces are part of the 'complex materialities of the urban' that pervade the city (Latham & McCormack, 2004: 703). Indeed, according to urban technology perspectives, the very coherence and intelligibility of the city as a social, economic, cultural and political organizational unit is thoroughly tied up with its technologies (Bulkeley et al., 2010; Hodson & Marvin, 2010; Smith, 2010). When it comes to understanding the dynamics of technological transformation of the city, a strong theme that emerges from these approaches is that new urban technologies will not mechanically lead to utopian sustainable cities that are more carbon neutral, more just and more liveable for all. Indeed, urban technology scholars are wary of straightforwardly celebrating technological innovation (Hughes, 1983; Gandy, 2005; Heynen et al., 2006; Swyngedouw, 2006).

With regards to innovation in energy technologies specifically, Coutard and Rutherford (2010) argue that that the relationship between technology and the city is a priori ambivalent and likely to be socio-spatially uneven and heterogeneous (see also Graham & Marvin, 2001; Coutard & Guy, 2007). In this context, interpretations of infrastructural change that see energy infrastructures as key 'for climate protection and the whole ecological modernization of societies' (Monstadt, 2007: 327) stand in stark contrast to Graham and Marvin's (2001) dystopian portrayal of a splintering urbanism, where technologies are implicated in the de- and re-bundling of parts of the city into premium spaces while others are bypassed by these trends, partially disconnected or completely 'dumped'. Critical urban scholarship draws attention to the 'power relations and regulatory pressures within which sustainable cities are being forged . . . and who benefits most from these strategic formulations' (Whitehead, 2003: 1192; Hodson & Marvin, 2009a). Rather than a wholly novel area of concern to urban governance, political aspirations of ecologically and socially sustainable cities are cast as a particular way of reframing the management of urban areas (Gibbs & Jonas, 2000: 355) – a way in which urban elites may 'speak on behalf of the city' (Hodson & Marvin, 2012), possibly at the risk of silencing more marginal voices.

Viewed thus, the reconfiguring of the city can be understood as a highly normative enterprise that is intimately tied up with particular ideas about progress and 'the good'. As the example of Barcelona's iconic urban PV power station suggests, urban change is rarely uncontested and straightforward. As pre-existing, highly material and social environments, cities may exhibit resistance to innovative forces. The challenge of introducing new technologies into existing arrangements is that these may preclude the formation of alternative possibilities. Urban technology scholar Anique Hommels (2005) proposes that one must be attuned to the nature of urban obduracy – place-contingent configurations that precede new technologies – in order to understand the place-specific 'interplay of a variety of social, technical, cultural, and economic factors' (Hommels, 2005: 328). Also theorized as regimes in the broader innovation literature, existing configurations of hardware and social relations are frequently understood as the dominant grammar, or the context in which innovation takes shape (e.g. Rip & Kemp, 1998; Smith et al., 2005). Existing incumbent technologies, actors, institutions and elites dominate the status quo, while attempts to innovate face the challenge that novelty is frequently marginalized in existing practices. Within Western energy systems, this has been commonly theorized as a carbon lock-in (Unruh, 2000). Here, the contemporary landscape of large and centralized fossil fuel power stations is understood as the product of historical system-building processes that have gained momentum due to large sunk investments, economies of scale and powerful, institutionalized elites. As the backdrop providing stability and durability to existing material and social relationships around energy, these mean that new energy technologies face structural disadvantages that can only be undone at high cost and effort.

Yet, in normatively-driven processes of urban transformation, new technologies become significant as crucial material nexuses for turning a particular conception of the desirable future into actual urban spaces that gather novel and more desirable behaviours. Conceptually, this has been portrayed as a form of spatial politics: a practice of converting aspirations and visions into material outcomes, turning 'the nowhere of the imagination into the good place' (St John, 1999: 55). For example, the urban PV power station at the Forum of the Cultures in Barcelona – while clearly not uncontroversial – can be read as a script (Akrich, 1992: 208) of the Ajuntament's aspirations for Barcelona's new energy future around which urban energy relationships may take hold (Figure 9.1). Such utopian interventionist spatial politics has been theorized by a range of authors using different terminology, including as 'utopics' by Marin (1984) and Hetherington (1997), and as 'modes of mattering' by Law (2004). What is helpful about the work of these scholars is that it brings into focus the role of urban technologies in converting the aspirations of (a potential diversity) of actors into actual urban space. Scholars sensitive to the role of material objects in reconfiguring practices and policies (e.g. Shove, 2003; Lovell, 2005) have explored the role of things like shower timers (Hobson, 2006)

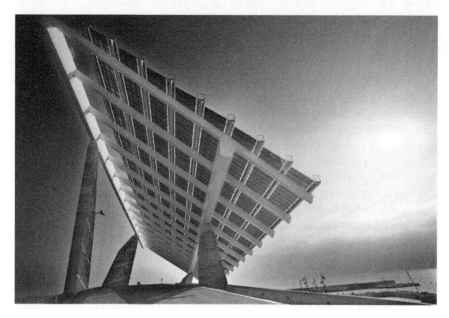

Figure 9.1 Photovoltaic Pergola in the *Forum de las Culturas*, Barcelona
Source: Isofoton (Creative Commons Attribution 3.0)

in changing water usage and solar panels in shaping energy consumption behaviours (Keirstead, 2006; Abi Ghanem & Haggett, 2011). Lovell (2005: 2500), for instance, notes how housing materials and other technologies are the 'substance of policy' that critically enable (though do not determine) the emergence of different eco-housing policy storylines and the formation of alternative discourse coalitions. In these subtle yet fundamental ways, technologies may serve to mediate the possibility of change.

Notions surrounding the obduracies of existing configurations and the importance of urban technologies in enabling the formation of new possibilities provide useful conceptual foundations for thinking about the prospects of innovation in urban energy systems. While solar technologies may not be straightforwardly beneficial, their very existence enables innovators to spatialize their aspirations for new policies, behaviours and practices around urban energy. This chapter now turns to charting the dynamics underlying the rise to prominence of solar technologies in Barcelona between 1998 and 2010. Just like the historical fate of solar technologies more generally has been tied up with the geopolitics of the twentieth century in terms of inter-state competition, economic development and emerging global ecological concerns (see Perlin, 2000), Barcelona's solar story is replete with simultaneously socio-economic and politically-charged relationships around urban energy. To put the events described in this chapter into context, the next section traces several trends at local, national and international scales which created the conditions for solar technologies to assume significance in Barcelona around the turn of the millennium.

Solar Cities Gather Momentum

It may appear surprising at first glance that solar technologies came to assume significance in Barcelona in the period between 1998 and 2010. Historically a major industrial centre, until the early to mid-1990s there were few signs of the city's environmental aspirations. An obdurate, de-industrializing environment, Barcelona is embedded into a regionally centralized energy system with strong utilities and a predominantly nuclear generation base, due to how the electricity sector was carved up between regional economic elites following the Spanish Civil War (1936–39). In addition, Barcelona is an old compact city with limited space, an aging building stock (that is renewing only very slowly), and a built environment that is structurally varied and diffuse in ownership and management arrangements. Significantly, solar technologies are not quintessentially urban technologies. Their historical niche in off-grid applications, such as low-performance electric consumer goods, oil rigs, emergency call boxes and remote rural settlements, place them away from the urban. In the European context, cities are the historical birthplace of electricity grids and constitute perhaps the most grid-connected places on earth. Bringing solar technologies to the city implies 'solarising the electrified' (Perlin, 2000). At the time of writing this chapter, solar technologies are characterized by high upfront capital investment costs which adversely shape the economics and financial viability of these technologies.

Yet, despite a number of adverse factors, there was a range of favourable developments that generated a climate conducive to urban-scale solar energy at the local level. At a global scale, the issue of urban sustainability had come to assume prominence in 1992, largely as a result of the United Nations Conference on Environment and Development (UNCED), often referred to as the Rio de Janeiro Earth Summit. Public policy audiences became increasingly interested in exploring the role of cities in the context of climate change mitigation – given that they are major emitters of greenhouse gases based on their concentrated energy consumption patterns. However, while a critical site for climate policy, a first wave of sustainable cities and urban climate governance researchers warned that the ambitions of municipal authorities to reduce emissions are frequently at odds with their sheer resource intensity and dependence on out of town energy supplies (Collier, 1997; Haughton, 1997; Nijkamp & Pepping, 1998; Bulkeley & Betsill, 2005; Bulkeley & Kern, 2006). For cities, in the absence of being able to control the source of supply, decreasing their reliance on out of town power stations was thought to be a salient way to exercise control over the environmental quality of their energy supply, and thus for evading the carbon lock-in of the wider energy system. In this context, technological advances in decentralized energy technologies, including solar energy, became increasingly seen as critical to 'internalising' a proportion of energy supply within the administrative urban boundary, making place for cities that are more ecologically sustainable (Bulkeley & Schroeder,

2008; Hodson & Marvin, 2009b). Within Europe, while the green credentials of alternative energy technologies had been the focus of civil society groups in existence in the 1970s (e.g. the Alternative Technology Movement; see Smith, 2003), the drive to decarbonize was based on growing acceptance of the scientific theory of human-induced climate change. This underpinned the proliferation of national renewable energy strategies and public financial subsidies during the late 1990s. For instance, at the level of the Spanish state, the seminal *Real Decreto*, or Spain's executive decree 2818/1998, legalized feed-in tariffs, establishing for the first time the legal right of renewable energy producers to sell their entire electricity production to the grid at a premium above the wholesale market price. Besides an environmental necessity, solar technologies became increasingly positioned as an economic opportunity for diversifying the economic base of the national economy, then dominated by a housing boom in the south of the country. Further legitimizing renewable energy technologies, the 1999 Spanish renewable energy strategy declared that the national PV sector 'constitutes an important reservoir of employment and an opportunity for business creation and the development of new industrial sectors' (Gobierno de España, 1999: 16).

In Barcelona, Spain's second city and the regional capital of the autonomous community of Catalonia, solar technologies resonated strongly with the Agenda 21 movement and its principles of resilience, liveability and productivity. The city's energy consumption had been steadily increasing, along with its population, rising living standards and the growth of the service sector. While the city's energy mix was predominantly nuclear (49 per cent), Barcelona's CO_2 emissions increased by 10.78 per cent between 1999 and 2006; in 2010, over 73 per cent of emissions resulted from the city's energy use (Maassen, 2010).[1] The need for urban sustainability was seen with increasing urgency, particularly in the context of European-scale processes of tightening environmental standards (Graham & Guy, 1995; Guy et al., 1996; Moss et al., 2001). Benefiting from this momentum, a critical mass at the city level emerged that positioned sustainability issues as central to urban development. Following the 1998 local elections, where the Green Party (*Ecologistes en Acció*) gained control over the municipal council, the city established a multi-stakeholder and cross-sectoral Environment and Sustainability Council, which initiated one of the first participatory urban Agenda 21 processes worldwide. By the mid-2000s the Ajuntament was a signatory to a number of international agreements and movements on climate change. By 2010 Barcelona had become a celebrated poster child for European climate and environmental lobby organizations such as ICLEI,[2] PV Upscale[3] and Energy-Cities.[4]

Municipal Engineers and Solar Guerrillas

To understand how solar energy came to assume significance in Barcelona, it is essential to trace the activities of two sets of protagonists. These, for the purpose of this chapter, will be referred to as the 'municipal engineers', on the one hand, and the 'solar guerrillas' on the other. The municipal engineers were instrumental in the design and delivery of the seminal urban energy policies and initiatives for which Barcelona would acquire national and international fame. Their very existence was the outcome of a series of political decisions that created the Barcelona Energy Agency (AEB). This was established in 2002, when the Ajuntament proclaimed Barcelona's 'new energy culture'. The AEB became a municipal entity staffed predominantly with engineers and tasked with executing the logistical side of the municipal energy policy. The AEB's municipal engineers supported the implementation of the trail-blazing environmental planning by-law, the Solar Thermal Ordinance (approved previously, in 2000), establishing the requirement for new buildings and those undergoing major renovation to meet at least 60 per cent of their demand using solar thermal energy. Other activities included awareness raising campaigns (e.g. 'Barcelona Saves Energy') and demonstration projects such as solar thermal installations in public swimming pools and solar photovoltaic systems in schools, community centres and other public buildings.

In parallel to the genesis of the municipal engineers, a set of activists began grouping around the 'Solar Guerrilla Manifesto', the charter of a local Barcelona-based non-governmental organization (NGO) called the Fundació Terra. The Manifesto empowers 'the solar guerrillas of planet earth' to advance a sustainable energy strategy by being

> . . . exemplary in our households by generating electricity using clean and renewables sources such as the sun, the wind, water or biomass. . . . We, the solar guerrillas of planet earth, are committed to use safe and certified technologies that do not harm our neighbours, the workers in the energy sector or our environment . . . above all, the solar guerrillas are a movement of civil awareness and compromise with future generations.
>
> (Fundació Terra, undated)

Influenced by the Rio Earth Summit, the NGO began operating in 1998, staffed with a set of individuals who had sympathies with the same Green Party that had secured influence during the local elections. Initially, this solar guerrilla organization expended much of its energy on involving civil society in renewable energy initiatives through educational and leisure campaigns. More far-reaching, however, were some of the Fundació's more transgressive activities. The NGO pioneered small-scale solar power at a time when non-grid utility electricity generation was still largely unheard of in Spain. A vocal critic of the government's renewable energy policy, the solar guerrillas were always in search of new ways of unsettling deeply entrenched relationships around energy by targeting the powers that be and civil society more generally.

As examined in the remainder of this section, both protagonists in Barcelona's energy transformation positioned themselves against the status quo of centralized energy generation, which made Barcelona dependent on out of town supplies. However, it was not until engineers and guerrillas began converting their aspirations for non-centralized urban energy into actual urban spaces that the implications of their different ambitions, origins and capabilities became apparent and meaningful.

Exposing the Status Quo

Despite their rather different origins and backgrounds, both the municipal engineers and the solar guerrillas shared a basic dissatisfaction with Barcelona's status quo – its energy dependence on centralized nuclear and fossil fuel-based grid energy, which reduced the city to a passive energy consumer. Further in common was that both protagonists began exposing this status quo by generating new forms of knowledge to generate evidence for action. On the one hand, the municipal engineers' first task, following the creation of the AEB, was to produce the Barcelona Energy Improvement Plan – also known as the PMEB, or *Pla de Millora Energètica de Barcelonai* (Ajuntament de Barcelona, 2003), which would become the cornerstone of the Ajuntament's renewable energy strategy. This document, the first of its kind in Barcelona, Spain and beyond, provided a sophisticated technical analysis of urban energy flows, a scenario-based projection and an action plan. Barcelona's solar guerrillas also generated novel forms of understanding the city as space for power generation. Through the Fundació, they hosted large *solar paellas* and *solar chocolates*, public cooking events designed to position solar power as a real possibility in Barcelona. Further, analogously to the PMEB, members of the Fundació, alongside one of Barcelona's prominent solar guerrillas, produced a study showing that if the rooftop surfaces in one specific Barcelonan district (the *Eixample*, a district characterized by regular architectural form, geographical layout and relatively well-known rooftop usage) were covered with photovoltaic panels, over 60 per cent of domestic electricity consumption could be met by solar electric technologies (Bosque Garcia & Domingo Marín, 2008).

Through quantitatively dissecting the city's energy flows into constituent components and demonstrating the power of the sun, engineers and guerrillas were enabling Barcelona's buildings, public space and transport infrastructures to become reinterpreted as potential sites of solar electricity generation. The calculations and maps of energy flows they produced portrayed Barcelona in novel and unusual ways – less concerned with geographical physical and territorial boundaries than with the interconnection of energy flows across space. The evidence this generated – about Barcelona's high dependence on external energy supply – exposed the status quo and paved the way for the range of subsequent activities by which engineers and guerrillas would turn their solar aspirations into actual urban space. Yet, while sharing in common an opposition to the status quo of centralized electricity generation, it is precisely in the different ways in which municipal engineers and the solar guerrillas went about spatializing their aspirations for alternative futures that meaningful differences became apparent between the politics of their respective ambitions.

Transgressing the Grid

The notion that renewable energy technologies are locked out from existing energy systems (Kemp, 1994; Unruh, 2000; Del Rio & Unruh, 2007) is a useful starting point for describing the challenges encountered by the solar guerrillas in the early days following the coming into force of the national Spanish feed-in tariff. When the Spanish government introduced the first feed-in tariff in 1998, the Fundació Terra pioneered the use of decentralized small-scale solar power both in Barcelona and Spain. This was taking place at a time when non-grid utility electricity generation was still largely unheard of in Spain – in contrast, the country's emerging solar capacity was taking the shape of large-scale solar installations, the so-called solar farms, located outside urban boundaries. The President of Fundació Terra, a self-confessed solar guerrilla, explained what happened when they sought connection to the national grid for their small-scale (2.2 kWp) terrace photovoltaic installation:

> ... back then it was not envisaged by the regulation that small-scale connections in monophasic [current] would be made, but only in triphasic [current]. But not in any household you will find triphasic. And neither in an office, obviously. That's for industry and engines.
>
> (Interview, 2009)

Upon seeking connection, the Fundació promptly came up against the dominant electricity generation logic at the time. The issue was one of the novel scale of power generation – the statement above refers to the different ways in which power is transformed as it is transmitted over distance. Here the use of 'monophasic' and 'triphasic' indicates different grid setups, where monophasic refers to local grids which domestic/urban users are connected to whilst triphasic current refers to the connection used in Spain for industrial consumers. Indeed, it was not anticipated by the *Real Decreto* that small-scale grid connections would be made, as power generation was implicitly thought to be a large-scale matter. The dominant type of solar power anticipated in Spain was the large-scale (and non-urban) solar farm.

By seeking connection to the grid for their small installation, the Fundació set a technical and legal precedent with the local utility's engineers and appealed to the Spanish High Court to denounce the substantial administrative barriers faced by small-scale power generation in Spain. At the time, the *Real Decreto* framework required individual generators to keep a record of all magnitudes of electricity generated, to issue monthly bills to the utility and to fill out a tri-annual tax return – even though the power produced was taxed at zero per cent, non-compliance meant facing a financial penalty. The amount of paperwork involved to access the national feed-in tariff under the *Real Decreto* was described by an energy advisor from a local energy non-profit as 'a dealbreaker' (Interview, 2009). The solar guerrillas, through the Fundació, filed a legal complaint to the Spanish High Court of Justice concerning the 'absurdity' of the policy, based on the contradiction between its ostensible framing as a mechanism geared towards rewarding small-scale generation (with an initial production ceiling of a 5 kW peak to qualify for the feed-in tariff), and the reality of the challenges facing the grid-connection of such small-scale systems. The outcome of the legal proceedings demonstrated what is described by a solar advocate as a 'veritable obstacle course' constituted by an opaque distribution of governmental competences and powerful utilities:

> The judgment of the high court of justice arrived in 2004, many years later. The decision acknowledged that we were of course entirely correct, but that this was a problem of the legislator and it was not the tribunal's role to judge what the legislator does . . . it won't change, not for the time being, the utilities are scared that small-scale power gathers, let's say, momentum.
>
> (Interview, 2009)

Others expressed similar opinions about the incumbent utility's perception of the threat of the impact of decentralization on revenues, and the unspoken alliance between powerful economic interests across the energy and construction sectors – for instance, that big construction firms 'are practically in bed with the utilities . . . a very incestuous group' (Interview, 2009). Faced with the unsuccessful attempts at re-articulating the socio-spatial distribution of energy relationships and failure to fully legitimize the household as a new site for power generation, the solar guerrillas declared a 'solar strike', which consisted of disconnecting their system from the wider grid.[5]

Going forward, the solar guerrillas developed a portfolio of activities which was characterized by different degrees of activism. The *Guerrilla Solar* campaign constituted the Fundació's most transgressive form of activism, where solar guerrilla refers specifically to a person who connects a small-scale solar power generator to feed electricity back into the public electricity grid without the permission or knowledge of the grid owner. The campaign consisted of the promotion of a small PV panel (the photonic kit GS120) that users simply plug into their electricity sockets – but instead of consuming energy the panels generate electricity (about the amount consumed by a common household fridge). Crucially, this sort of plug-and-play mode of generation, by which a PV is simply connected to the grid without prior authorization from the utility, was not legal in Spain; neither, however, was it illegal. The sub-kilowatt capacity of the photonic kit meant that it did not qualify as an electricity generator in legal terms. The solar guerrillas thus took advantage of this legal loophole by which the sort of micro-scale electricity generated by the guerrilla panels was situated in a grey area. The *Ola Solar* (Solar Wave) was another way in which the solar guerrillas sought to increase citizens' participation in renewable energy. The *Ola Solar* was a one-off multi-investor participatory PV installation (of 43.7 kWp) on one of Barcelona's markets, which brought together 140 individual 'investors'. Each contributed between €1,000 and €3,000 to the systems' costs and each henceforth received yearly dividends of roughly €100 from the income generated by the sale of electricity to the utility under the Spanish feed-in tariff.

What became clear through the transgressive activities of the solar guerrillas is how their attempts to spatialize their aspirations for greater citizen participation in energy generation – and thus transforming established relationships around energy – faced considerable obstacles in a deeply entrenched energy system such as the Spanish one. The scale of generation and, crucially, ownership patterns were strongly resistant to attempts to bring about greater public participation in energy generation. Meanwhile, the municipal engineers, seeking to execute the progressive energy policies of the Ajuntament, were also coming up against the obduracies of the status quo. In their case, however, converting municipal ambitions into urban spaces faced the challenge of re-engineering existing public buildings to be compatible with the ambition of reducing municipal carbon emissions.

Un-engineering Urban Obduracy

The process of approving the city's solar thermal planning by-law, which targeted the installation of solar energy systems on new buildings, was largely a smooth process based on a consensus between the governing coalition in the municipal government and the local NGO lobby (especially the energy agency BarnaGEL). However, converting the Ajuntament's plans for solar photovoltaic installations into actual solar spaces on existing buildings came up against what Hommels (2005, 2008) terms 'urban obduracy' – the fixity of the city, as it is securely anchored in its own history. Photovoltaic demonstration projects in municipal buildings began in 2000 with installations on the roofs of the Ajuntament's two main administrative buildings, Nou and Novissim, in central Barcelona. Subsequently, over 30 other PV installations were placed across the city's ten districts, focusing on community centres for the young and old, schools and other public and civic buildings. Yet, introducing new energy technologies into the building stock required challenging existing urban planning modes, which proved resistant to the new imperatives of the times. Specifically, re-engineering public buildings for solar energy would require a new order of urban planners, architects, roofers and electricians with the willingness as well as – crucially – the know-how to match calculations of structural loads and dimensions of buildings and rooftops (conceptualized in terms of weights and surface areas) with energetic considerations (expressed in kilowatts and kilowatt-hours).

In Barcelona most power in the field of urban planning lies with the Ajuntament. However, the city's ten administrative districts have some powers in the area of planning and infrastructure. When the Ajuntament put the municipal engineers in charge of delivering a series of PV installations on municipal buildings in 2002, they encountered significant problems with securing the cooperation of the suburban districts' Technical Services. These professionals, in charge of the maintenance of public buildings at the local district level, are predominantly built environment professionals with architecture backgrounds (in contrast to the municipal engineers who are well-versed in new energy technologies). The following quote from a municipal engineer involved in the solar projects explains how delivering the PV installations was a 'tough fight' with the architects from the Technical Services:

. . . the [district] architects were telling us, 'no inclination on the panels!'. But panels without inclination, apart from being less efficient, also do not clean themselves. What did we do? Well, it was a fairly tough fight in this regard! . . . The majority of architects still live in the era of pyramids [laughs], where the only thing they do is put stone upon stone. . . . It's a problem of training. On the curriculum of architects renewables do not exist.

(Interview, 2009)

Effectively acting as gatekeepers to the local public building stock, the Technical Services' reluctance to engage with the new technology stalled the delivery of the PV projects. Several attempts to transform district-level planning systems were made over the course of numerous installations across the city. According to the account of the municipal engineers, this process was marked by some degree of confrontation and substantial technical learning. Eventually, experience gained from experimentation led the Technical Services architects, over a period of several years, to align with the municipal engineers' intention of installing PV in municipal buildings. As recounted by another municipal engineer:

It was difficult at first, coming in with this [solar PV] . . . but the evolution in all the districts has been considerable. There's awareness, there's knowledge. Now they know the topic of PV and they are starting to know the structures that enable integrating PV into buildings . . . when there's a new building they call us and say 'listen, we'll prepare the building so you can put some photovoltaic panels'.

(Interview, 2009)

The cumulative outcomes of the municipal learning processes described above were such that the municipal engineers were able to place solar installations on a number of public buildings, including numerous schools across the city's ten districts, effectively converting ambitions for municipal solar energy into real urban spaces. Solar schools in particular have come to be presented as tangible spaces in which children's everyday encounters with energy are opened up for negotiation and reconfiguration and may effect generational changes in engagements with energy. Solar schools in Barcelona, as elsewhere, feature as a critical educational device for the pupils and their families, as well as for the district architects who have become familiar (and even enthusiastic) about experimenting with different forms of building integration. Through the public buildings campaign, the municipal engineers were eventually able to generate sufficient support to instate a solar PV planning policy that requires all new public and private developments and renovations above a certain size to install a certain amount of PV power (effectively constituting a follow-up from Barcelona's pioneering Solar Thermal Ordinance).[6] Step by step, the Ajuntament, supported by the energy-sophisticated municipal engineers, was realizing its ambition of turning Barcelona into one of the cities making 'the most use of solar energy' (Ajuntament de Barcelona, 2003: 61).

Confronting the Politics of Divergent Ambitions

Both municipal engineers and the solar guerrillas challenged the status quo and came up against pre-existing configurations that were commensurate to their respective ambitions and entry points – competing jurisdictions over urban space and grid politics, respectively. In this way, both protagonists' attempts to carve out new urban spaces for materializing their alternative ambitions highlight that processes of reconfiguring existing urban energy relationships can be highly contested. However, in addition to drawing out the obduracies of the status quo, the case of the municipal engineers and the solar guerrillas brings into focus the fault lines that can emerge between different innovators' attempts to reconfigure the status quo according to their divergent ambitions. This section explores several instances that emphasize the confrontations that can occur when divergent spatial politics collide – as was the case with the municipal engineers' ambition of municipal carbon reduction and solar guerrillas' eco-empowerment aspirations.

While the municipal engineers targeted urban sites largely within their reach, such as public buildings over which they had some jurisdiction, the solar guerrillas aimed for a much more far-reaching transformation. The latter's ambitions transcended the urban, targeting regional and national spheres of the energy system. Significantly, instead of using solar energy for achieving municipal carbon reductions, visibility and international standing, the solar guerrillas targeted the environmental implications of the entire socio-economic organization of the electricity sector in Spain. In this way, the solar guerrillas' version of eco-empowerment was explicit in targeting all traditionally passive sites of consumption – be they domestic, community and other. In contrast, the municipal authority's narrower ambition to reduce carbon was much less discriminating of what urban space was the target for intervention, mainly the 30 or so photovoltaic solar energy installations on public and civic buildings and other public spaces throughout the city's ten districts. Indeed, the visions and strategies of the Ajuntament's municipal engineers and the solar guerrillas of Fundació Terra were not always readily compatible.

The latter's push for more ambitious policy commitments at all levels was overtly not aligned to the municipal engineers' exclusive focus on municipal demonstration projects and securing Barcelona's 'high position in the international urban ranking' (Ajuntament de Barcelona, 2003). Attempting to turn clients into generators, the Fundació Terra foresaw a much greater role for solar energy to be played in Barcelona (Fundació Terra, 2007; Bosque Garcia & Domingo Marín, 2008) than anticipated by the municipal demonstration projects. Embedded in the solar guerrillas' 'ecological activist' beliefs (see Fundació Terra, 2014) was a much more far-reaching vision of eco-empowerment, consisting of socio-environmental emancipation through taking responsibility for one's consumption. Crucially, one which contained a vocal critique of public institutions' perceived apathy:

> the strategic plans at the autonomous and municipal scale, of Catalonia and Barcelona, respectively, are extremely conservative. . . . We believe that

Spain should become a role model in energy matters for other countries to follow, and Catalonia should not overlook this opportunity to become one of the pioneering autonomous communities.

(Bosque Garcia & Domingo Marín, 2008: 78)

The ambitions of the solar guerrillas were for the urban (as well as regional and national) authority to take on a much more proactive role in the promotion of solar energy. Thus, in parallel to urban-scale activism, the solar guerrillas through Fundació Terra worked with Greenpeace's Spanish arm and the pan-European non-profit European Association for Renewable Energy (Eurosolar) to push their agenda. Beyond the confines of city politics, they promoted a 100 per cent renewable energy system for both Catalonia and Spain (Fundació Terra, 2007; Greenpeace España, 2007). For activists such as Josep 'Pep' Puig, who guided the Fundació Terra's analysis of the *Eixample* and himself a key protagonist of the solar guerrilla movement, it was a matter of pushing for a greater political ambition – 'because you can enact an ambitious politics, or one that is more conservative. And normally public administrations here do the latter' (Interview, 2009).

The multi-investor project *Ola Solar,* the Guerrilla panel and their own small terrace system, along with their various other initiatives, are emblematic of the solar guerrillas' aspirations of broadening participation in renewable energy whilst putting pressure on the Spanish government for regulatory change. All of the solar guerrillas' initiatives transgress citizens' established (dis)engagement from energy infrastructures. The underlying conviction, embedded in the solar guerrilla manifesto, is that involving regular citizens in energy generation unsettles both the environmental equation in existing energy systems as well as, crucially, the unequal client relationship between citizen and corporate interests. The intention behind the *Ola Solar* specifically was to turn citizens into 'shareholders of the sun' and an otherwise passively energy-consuming municipal market into an 'initiative of popular capitalism' (Fundació Terra, 2007). The President of the Fundació Terra, a solar guerrilla himself, explains how scaling up this initiative, to reconfigure even more regular citizens into power generators, was not one that was shared with the Ajuntament:

> You had the Ajuntament, doing their own installations on public buildings. . . .
> We've always argued that this is not the best way to do it, to make solar available to more people. . . . We designed the *Ola Solar* . . . and we proposed ourselves [for managing further participatory projects] . . . but they wouldn't let us. It was too complicated, because if you are the Ajuntament, and you just want to do it, then you do without the hassle of negotiating with an NGO.
> (Interview, 2009)

The municipal engineers evidently did not (as members of a relatively larger organizational unit) struggle with the implications of the convoluted regulatory system in the same way as households would. Nor were they concerned with making a principled statement about the treatment of small-scale solar electricity generation in Spain, despite their self-positioning as key players in stimulating the emergence of greener alternatives. Because of their relative lack of political engagement with exposing and undoing the systemic bias against non-centralized power generation in Spain (contrary to the solar guerrillas' ambitions), activists perceived the urban authority as more concerned with being 'seen to be green' for the purpose of its urban marketing strategy. They rejected what they saw as a unilateral approach focusing on appearances – to not fall behind in carbon emission reduction 'rankings' and outperforming other cities.

According to some of the solar guerrillas, the iconic urban PV power station at the 'Forum of the Cultures' that introduced this chapter was emblematic of this municipal unilateralism and imperviousness. While fundamentally sharing a belief in the importance of visible sustainability projects and leadership initiatives, for its opponents the installation embodied the municipality's strategic use of solar energy in public spaces for promotional rather than eco-empowerment purposes. Positioned as part of Barcelona's new ecologically friendly urban 'energy culture' (Ajuntament de Barcelona, 2003), the installation was a site of controversy from the moment of its unveiling. During the 2004 Universal Forum of Cultures event, organized by the Ajuntament, the installation was the site of civil protests in which citizens took a stance against the event's business orientation. In particular, they protested against the dubious reputation of the event sponsors and speakers (including several with questionable trade practices and policies in the developing world); and the fact (interpreted as hypocrisy) that the event organizers did not condemn the 2003 invasion of Iraq, which was thought to be related to several sponsors of the event having stakes in the arms industry. Interpretations of the installation as an icon of sustainability thus stood in stark contrast to citizens reclaiming the Forum 2004 area on Barcelona's esplanade as a 'peoples' space', as the event had intended it. In this way, the protests added to existing questions concerned with the kind of a leader Barcelona ought to be – in solar energy, as well as more generally.

Technology, Innovation Dynamics and Prospects of Re-inventing the City

The context of the research underlying this chapter (Maassen, 2012) was a question of whether and how photovoltaic (PV) technology has come not only 'from space to earth' (Perlin, 2000) but also to the city, given growing interest in technology-related urban transformation for sustainability. In Barcelona, a number of actors came to formulate their aspirations about a desirable future state of affairs using solar energy – specifically, seeking to deviate from the existing setup of fossil fuel and nuclear-based centralized urban energy provision, by which energy was imported into the city from out of town power stations. Given the obduracy of pre-existing relations that constrain the potential to reconfigure the urban, the carving out of spaces for renewable and decentralized energy technologies was a very real achievement in a city as dense as Barcelona and a grid setup as restrictive as the

Catalan. The chapter provided an exploration of two types of urban-based actors who were seeking to promote urban-scale solar technologies: the solar guerrillas, a set of non-governmental activists, and the municipal engineers, charged by the local authority to realize solar installations across the city. These protagonists pursued rather different innovation strategies, based on their different capabilities as well as solar visions that diverged in ambition, scope and political orientation. As such, the case of Barcelona's municipal engineers and the solar guerrillas sheds light onto the role and nature of technology, processes of innovation and the potential transformational effects of innovation for urban energy systems – which are salient questions in contemporary contexts of climate change, resource constraints and social equity.

In the first instance, solar technology became meaningful as a practical material and discursive locus around which the chapter's two central protagonists, the municipal engineers and the solar guerrillas, sought to spatialize their aspirations about a desirable future state of affairs. However, while both were effectively seeking to alter an undesirable status quo, divergent visions and strategies generated a range of novel, alternative urban spaces. These spaces – urban icons such as the forum pergola, the participatory municipal projects and the solar guerrilla panels – were designed to embody the ideals of their creators. Once in place, however, they became open to contradictory interpretations, which foregrounded the protagonists' different visions of sustainability that diverged in ambition, scope and political orientation. Indeed, the aspirations for sustainability of the solar guerrillas and the Ajuntament's municipal engineers were not based on shared or converging visions, and nor were they readily compatible. While sharing elements in common, for the Ajuntament it was the low carbon international standing of the city that was of concern; from the activists' perspective this approach lacked a more fundamental interrogation of the nature of existing power relations and interests in the energy sector. Thus, in the process of innovating, new technologies such as solar energy in Barcelona become, simultaneously, vehicles for, as well as outcomes of, different actors' attempts to bring forth into the present aspirations for alternative futures.

In the second instance, differences in ambition, scope and political orientation draw attention to the way in which innovation unfolds across a diversity of sites. Such differences necessarily involve challenging a myriad of simultaneously social, material, spatial and temporal relations that make up the status quo. Visions and strategies may, for instance, stretch into the territory of national legislation (as with the challenge enacted by the PV system placed at the Fundació Terra's terrace) or extend to other cities in Europe (as in the case of municipal ambitions to outperform their urban European counterparts); they may involve suburban district professionals as well as regular citizens. A common theme across ambitions and strategies is the extent to which transforming obdurate environments through the introduction of novel technologies is a challenging endeavour. This realization goes right to the heart of the central question of urban innovation – the process and possibilities by which the boundaries between novelty and what is already in place (the 'context') are being redefined. In the extant literature, those material and discursive relations that predate efforts to innovate have been described as 'barriers' to action, socio-technical regimes and urban infrastructure regimes (Guy & Marvin, 1996; Shove, 1998; Guy & Shove, 2000; Bulkeley et al., 2005). While nuances in terminology indicate subtle shifts in conceptualizing the salient components of obdurate environments, notions of barriers and regime concepts consistently seek to capture the fact that innovation does not take place in a vacuum. Nowhere is this more pronounced than in cities, which are complex intersections of social and material relations that transcend territories, and human and infrastructural rhythms and rates of renewal.

In the third instance, to understand how technologies may transform cities, it is important to assess how successfully (technologies, and the relationships they gather) translate actors' visions into actual outcomes. The difficulty of reshaping the conditions of possibility, in the context of relations that predate the aspired-to states of affairs, is illustrated by the case of Barcelona: by 2010, solar installations were far from mushrooming in the city. Despite the eventual passing of the 'PV Ordinance' (making PV mandatory on certain new buildings and renovation projects), the policy spent five years in the pipeline (as opposed to its speedier counterpart, the Solar Thermal Ordinance). Its significance is drastically reduced by the fact that it is not thought to have a significant impact on a city as dense as Barcelona, where new constructions and renovations are few. The question of how to retrofit energy systems in existing buildings continues to be an ad hoc undertaking, with little suggestion that the rooftops of Barcelona's *Eixample*, or of any other Barcelonan district, will install solar power plants in the near future. While the solar guerrilla panels were a way of bypassing unfavourable national policies, the Spanish state's constant tinkering with the national feed-in tariffs has had deleterious consequences on the national PV industry in general. For the vast majority of prospective small-scale generators, implementing urban PV remains a perplexing endeavour owing to the convoluted administrative setup at the national scale. A solar guerrilla stated: 'there have been certain things going on here, a few years ago, but these have pretty much run out of steam' (Interview, 2009).

In a city such as Barcelona, with electricity demands in the tera- rather than giga-watt magnitudes, small-scale solar applications may seem negligible. Yet, several urban solar spaces came to matter in a more than a symbolic manner. The municipal demonstration projects engaged district-level technical professionals in a new way of thinking about the city. Through the solar guerrillas' pioneering installation and panels, legislative loopholes and grey areas were made intelligible, and the participatory PV installation and solar installation in schools have enabled more citizens than ever before to participate in shaping energy futures. There have been considerable changes, and for the first time urban space became re-imagined and demonstrated to be energetically valuable. This suggests that the electric possibilities that are latent in decentralized energy technologies make it possible to question and open up to contestation the fixity of boundaries between features such as cities, electricity systems and their interconnection by means of transmission lines. At the confluence of contradicting interpretations and aspirations, the repercussions of new technologies are felt far beyond their physical

boundaries – they indicate fundamental shifts in how urban space has traditionally been thought about, and treated in policy and practice. In Barcelona, solar
technologies were the basis from which actors began to deviate from and disorder
established relations of centralized energy geographies, transgressing traditional
production and consumption boundaries and questioning the legitimacy of
entrenched power relationships, commercial interests and governmental authorities at different scales as they competed over urban space. For better or worse, new
urban technologies emerge as important nexuses around which urban futures are
contested worldwide.

Notes

1 These figures refer to the City of Barcelona, which is the focus of this chapter and should
   not be confused with the Province of Barcelona (a larger entity with over 5 million
   inhabitants) or the Barcelona Metropolitan Area (approx. 3 million inhabitants), which
   is the sixth largest city-region in Europe, accounting for roughly 12.5 per cent of total
   Spanish GDP.
2 Local Governments for Sustainability (www.iclei.org).
3 An EU Intelligent Energy Europe initiative (http://www.pvupscale.org/).
4 The 'European Association of local authorities in energy transition' (www.energy-
   cities.eu).
5 To this day the system is not rewarded with the national feed-in tariff, but the system was
   in fact reconnected following the entry into force of the Kyoto Protocol, based on the
   foundation's principled aspiration to 'refund' some of the electricity it has drawn from
   the grid.
6 Approved in 2006, the PV Ordinance was in fact superseded the same year by the
   National Technical Building Code even before its coming into force (which was sched-
   uled for 2008). The effective legal document is the Royal Decree 314/2006 that approves
   the Technical Building Code (Real Decreto 314/2006 por el que se aprueba el Código
   Técnico de la Edificación).

References

Abi Ghanem, D. and C. Haggett, 2011. Shaping People's Engagement with Microgenera-
   tion Technology. In: P. Devine-Wright (ed.), *Renewable Energy and the Public*. London:
   Earthscan, pp. 149–66.
Agència d'Energia de Barcelona, 2003. Pla de Millora Energètica de Barcelona. [Online].
   Available from: https://w110.bcn.cat/MediAmbient/Continguts/Vectors_Ambientals/
   Agencia_de_lenergia/Documents/Fitxers/PMEB_integre_cat.pdf. [Accessed 27 January
   2016].
Ajuntament de Barcelona, 2003. *Pla de Millora Energètica de Barcelona*. Barcelona: Ajun-
   tament de Barcelona.
Akrich, M., 1992. The De-Scription of Technical Objects. In: W.E. Bijker and J. Law (eds.),
   Shaping Technology, Building Society: Studies in Sociotechnical Change. Cambridge:
   MIT Press, pp. 205–24.
Bosque Garcia, S. and N. Domingo Marín, 2008. *L'energia solar fotovoltaica com una
   alternativa en els espais urbans*. Masters Dissertation, Universitat Autònoma de Bar-
   celona. [Online]. Available from: http://www.recercat.cat/bitstream/handle/2072/5322/
   PFCBosqueDomingo.pdf?sequence=1. [Accessed: 27 January 2016].

Bulkeley, H. and M. Betsill, 2005. 'Rethinking Sustainable Cities: Multilevel Governance and the "Urban" Politics of Climate Change'. *Environmental Politics* 14(1): 42–63.

Bulkeley, H. and K. Kern, 2006. 'Local Government and the Governing of Climate Change in Germany and the UK'. *Urban Studies* 43(12): 2237–59.

Bulkeley, H. and H. Schroeder, 2008. *Governing Climate Change Post-2012: The Role of Global Cities-London*. Tyndall Centre for Climate Change Research Working Paper 123.

Bulkeley, H., M. Watson, R. Hudson and P. Weaver, 2005. 'Governing Municipal Waste: Towards a New Analytical Framework'. *Journal of Environmental Policy & Planning* 7(1): 1–23.

Bulkeley, H., V. Castán Broto and A. Maassen, 2010. Governing Urban Low Carbon Transitions. In: H. Bulkeley, V. Castán Broto, M. Hodson and S. Marvin (eds.), *Cities and Low Carbon Transitions*. London: Routledge, pp. 126–41.

Collier, U., 1997. 'Local Authorities and Climate Protection in the EU: Putting Subsidiarity into Practice?'. *Local Environment* 2(1): 39–57.

Coutard, O. and S. Guy, 2007. 'STS and the City: Politics and Practices of Hope'. *Science Technology & Human Values* 32(6): 713–34.

Coutard, O. and J. Rutherford, 2010. The Rise of Post-Networked Cities in Europe? Recombining Infrastructural, Ecological and Urban Transformations in Low Carbon Transitions. In: H. Bulkeley, V. Castán Broto, M. Hodson and S. Marvin (eds.), *Cities and Low Carbon Transitions*. London: Routledge, pp. 107–24.

Del Rio, P. and G.C. Unruh, 2007. 'Overcoming the Lock-out of Renewable Energy Technologies in Spain: The Cases of Wind and Solar Electricity'. *Renewable and Sustainable Energy Reviews* 11: 1498–513.

Fundació Terra, 2007. Catalunya Solar. El camino hacia un sistema eléctrico 100 % renovable. [Online]. Available from: http://www.ecoterra.org/es/publicaciones/publicaciones-varias/catalunya-solar-el-camino-hacia-un-sistema-electrico-100. [Accessed: 1 January 2012].

Fundació Terra, 2014. Website. [Online]. Available from: http://www.fundacionTerra.es/en/campaigns/simple-living-and-sustainability/guerrilla-solar-plainly-ecological-activism. [Accessed: 1 January 2012].

Fundació Terra, undated. Manifiesto de la Guerrilla Solar del Planeta Tierra. [Online]. Available from: http://www.fundaciontierra.es/es/campanas/decrecimiento-y-sostenibilidad/manifiesto-de-la-guerrilla-solar-del-planeta-tierra. [Accessed: 6 May 2015].

Gandy, M., 2005. 'Cyborg Urbanization: Complexity and Monstrosity in the Contemporary City'. *International Journal of Urban and Regional Research* 29(1): 26–49.

Gibbs, D.C. and A.E.G. Jonas, 2000. 'Governance and Regulation in Local Environmental Policy: The Utility of a Regime Approach'. *Geoforum* 31: 299–313.

Gobierno De España, 1999. *Plan De Fomento De Las Energías Renovables En España*. Barcelona: Ministerio de Ciencia y Tecnologia, Instituto para la Diversificación y Ahorro de la Energía (IDAE).

Graham, S. and S. Guy, 1995. 'Splintering Networks: Cities and Technical Networks in 1990s Britain'. *Electronic Working Paper No. 18*. School of Architecture, Planning & Landscape, Global Urban Research Unit & Centre for Urban Technology, Department of Town and Country Planning, University of Newcastle upon Tyne.

Graham, S. and S. Marvin, 2001. *Splintering Urbanism: Networked Infrastructures, Technological Mobilities and the Urban Condition*. London: Routledge.

Greenpeace España, 2007. Renovables 100% Un sistema eléctrico renovable para la España peninsular y su viabilidad económica. Resumen de conclusiones. [Online]. Available from:

http://www.greenpeace.org/espana/Global/espana/report/cambio_climatico/resumen-conclusiones-100-reno.pdf. [Accessed: 27 January 2016].

Guy, S., S. Graham and S. Marvin, 1996. 'Privatized Utilities and Regional Governance: The New Regional Managers?'. *Regional Studies* 30(8): 733–9.

Guy, S. and S. Marvin, 1996. 'Disconnected Policy: The Shaping of Local Energy Management'. *Environment and Planning C-Government and Policy* 14(1): 145–58.

Guy, S. and E. Shove, 2000. *A Sociology of Energy, Buildings, and the Environment: Constructing Knowledge, Designing Practice*. London: Routledge.

Hård, M. and T.J. Misa, 2008. Modernizing European Cities: Technological Uniformity and Cultural Distinction. In: M. Hård and T.J. Misa (eds.), *Urban Machinery: Inside Modern European Cities*. Cambridge: The MIT Press, pp. 165–86.

Haughton, G., 1997. 'Environmental Justice and the Sustainable City'. *Journal of Planning Education and Research* 18(3): 233–43.

Hetherington, K., 1997. *The Badlands of Modernity: Heterotopia and Social Ordering*. London: Routledge.

Heynen, N., M. Kaika and E. Swyngedouw (eds.), 2006. *In the Nature of Cities: Urban Political Ecology and the Politics of Urban Metabolism*. London: Routledge.

Hobson, K., (2006). 'Bins, Bulbs and Shower Timers: On the "Techno-Ethics" of Sustainable Living'. *Ethics, Place and Environment* 9(3): 317–36.

Hodson, M. and S. Marvin, 2009a. 'Cities Mediating Technological Transitions: Understanding Visions, Intermediation and Consequences'. *Technology Analysis & Strategic Management* 21(4): 515–34.

Hodson, M. and S. Marvin, 2009b. '"Urban Ecological Security": A New Urban Paradigm?'. *International Journal of Urban and Regional Research* 33(1): 193–215.

Hodson, M. and S. Marvin, 2010. 'Can Cities Shape Socio-Technical Transitions and How Would We Know If They Were?'. *Research Policy* 39(4): 477–85.

Hodson, M. and S. Marvin, 2012. 'Mediating Low-Carbon Urban Transitions? Forms of Organisation, Knowledge and Action'. *European Planning Studies* 20(3): 421–39.

Hommels, A., 2005. 'Studying Obduracy in the City: Toward a Productive Fusion between Technology Studies and Urban Studies'. *Science Technology & Human Values* 30(3): 323–51.

Hommels, A., 2008. *Unbuilding Cities: Obduracy in Urban Sociotechnical Change*. Cambridge: MIT Press.

Hughes, T.P., 1983. *Networks of Power: Electrification in Western Society, 1880–1930*. Baltimore: Johns Hopkins University Press.

Keirstead, J., 2006. 'Behavioural Responses to Photovoltaic Systems in the UK Domestic Sector'. *Energy Policy* 35: 4128–41.

Kemp, R., 1994. 'Technology and the Transition to Environmental Sustainability: The Problem of Technological Regime Shifts'. *Futures* 26(10): 1023–46.

Latham, A. and D.P. McCormack, 2004. 'Moving Cities: Rethinking the Materialities of Urban Geographies'. *Progress in Human Geography* 28(6): 701–24.

Law, J., 2004. Matter-Ing: Or How Might STS Contribute? Centre for Science Studies, Lancaster University, Lancaster LA1 4YN, UK. [Online]. Available from: http://www.heterogeneities.net/publications/Law2004Matter-ing.pdf. [Accessed: 1 January 2012].

Lovell, H., 2005. *Low Energy Housing in the UK: Insights from a Science and Technology Studies Approach*. Department of Geography Durham, UK, Durham University. PhD.

Maassen, A., 2010. Barcelona. In: N. Cohen (ed.), *Green Cities (Volume 4) – The SAGE Reference Series*. SAGE Publications, Inc. [Online]. Available from: http://www.uk.sagepub.com/reference.nav. [Accessed: 1 January 2012].

Maassen, A., 2012. *Solar Cities in Europe: A Material Semiotic Analysis of Innovation in Urban Photovoltaics*. Department of Geography Durham, UK, Durham University. PhD.

Marin, L., 1984. *Utopics: Spatial Play*. London: Macmillan.

Moss, T., S. Guy and S. Marvin, (eds.), 2001. *Urban Infrastructure in Transition: Networks, Buildings and Plans*. London: Earthscan.

Nijkamp, P. and G. Pepping, 1998. 'A Meta-Analytical Evaluation of Sustainable City Initiatives'. *Urban Studies* 35(9): 1481–500.

Perlin, J., 2000. *From Space to Earth: The Story of Solar Electricity*. Cambridge: Harvard University Press.

PV Upscale, 2005. PV Upscale Strategies for the Development of PV in Barcelona. 30/06/2005. [Online]. Available from: http://www.pvupscale.org/IMG/pdf/Case-Study_Barcelona.pdf. [Accessed: 1 January 2012].

Rip, A. and R. Kemp, 1998. Technological Change. In: S. Rayner and L. Malone (eds.), *Human Choice and Climate Change, Vol 2 Resources and Technology*. Washington, DC: Batelle Press, pp. 327–91.

Shove, E., 1998. 'Gaps, Barriers and Conceptual Chasms: Theories of Technology Transfer and Energy in Buildings'. *Energy Policy* 26(15): 1105–12.

Shove, E., 2003. *Comfort, Cleanliness and Convenience: The Social Organization of Normality*. Oxford: Berg.

Smith, A., 2003. 'Transforming Technological Regimes for Sustainable Development: A Role for Alternative Technology Niches?'. *Science and Public Policy* 30(2): 127–35.

Smith, A., 2010. Community-Led Urban Transitions and Resilience: Performing Transition Towns in a City. In: H. Bulkeley, V. Castán Broto, M. Hodson, and S. Marvin (eds.), *Cities and Low Carbon Transitions*. London: Routledge, pp. 159–77.

Smith, A., A. Stirling, and F. Berkhout, 2005. 'The Governance of Sustainable Socio-Technical Transitions'. *Research Policy* 34(10): 1491–510.

St John, G., 1999. *Alternative Cultural Heterotopia: Confest as Australia's Marginal Centre*. School of Sociology, Politics and Anthropology, Faculty of Humanities and Social Sciences. Bundoora, Victoria (Australia), La Trobe University. PhD.

Star, S.L., 1999. 'The Ethnography of Infrastructure'. *American Behavioral Scientist* 43(3): 377–91.

Swyngedouw, E., 2006. 'Nature and the City. Making Environmental Policy in Toronto and Los Angeles'. *Annals of the Association of American Geographers* 96(1): 215–16.

Unruh, G.C., 2000. 'Understanding Carbon Lock-In'. *Energy Policy* 28(12): 817–30.

Whitehead, M., 2003. '(Re) Analysing the Sustainable City: Nature, Urbanisation and the Regulation of Socio-Environmental Relations in the UK'. *Urban Studies* 40(7): 1183–206.

10 Athens

Switching the Power Off, Turning the Power On – Urban Crisis and Emergent Protest Practices

Georgia Alexandri and Venetia Chatzi

Introduction

Since 2010 Greece has been portrayed as the black sheep of the European Union, held responsible for the current neoliberal crisis and the downward economic spiralling of the entire Southern European region. Under the pressure of the European Union (EU) and the International Monetary Fund (IMF), the country is facing ongoing neoliberal restructuring, which continues to impact the provision and consumption of public goods, including that of electricity. Neoliberal policies and associated tax mechanisms have led to increasing levels of deprivation, introducing novel forms of 'fuel poverty' (Bouzarovski, 2014) to the catalogue of social problems that cities like Athens have to face.

In Greece, electricity is generated, transmitted and distributed mainly by the state-owned Public Power Corporation, also known by its Greek acronym DEI. Official documentation refers to the company as a key public and social good (DEI, 2014a). Yet the way in which new austerity taxes on electricity consumption are being imposed disregards such public vision, paving the way for emerging forms of energy poverty and broader societal deprivation. Greece's neoliberal crisis finds an expression in the everyday through acute infrastructural crises, unsettling the social order and the urban experience and shifting how people relate to and use urban infrastructures (McFarlane & Rutherford, 2008). Yet, in Athens, austerity policies within the electricity sector are being challenged by new protest practices claiming a right to electricity provision free of the post-austerity taxes that are threatening this public good and deepening impoverishment. This chapter examines whether and how these novel practices of protest function as a 'crack' within a capitalist model (Holloway, 2010) which, within the context of the current neoliberal crisis, plays a key role in challenging the ongoing politics of austerity.

The chapter brings attention to the mobilisation within Greek society after the implementation, from 2010 onwards, of austerity measures and the taxation on the supply of power imposed by the creditors of the IMF, the EU and the European Central Bank (ECB). The first section recounts the crisis and austerity measures that led to social protests and mobilisation. Such mobilisation against energy taxation emerged in anti-austerity public meetings and assemblies, a social dynamic which resulted in the creation of new structures of thinking as well as spaces for discussion and collaboration amongst activists. Sections two and three focus on the protest practices that emerged in response to the taxation imposed on the consumption of electrical power (the Emergent Special Tax on Electrical Power Supplied to Built, popularly known as the Charatsi). Whilst activism and social networks achieved a postponement on the collection of the levy, the surveillance tactics of the neoliberal state ultimately suspended the impetus of these protests, which shifted their orientation towards the establishment of solidarity structures.

Crisis, Austerity Politics and Urban Uprisings

Neoliberalism is variegated in its forms rather than being a hegemonic force applied the same way in all geographical contexts (Peck et al., 2009). Crisis is structurally linked to the capitalist mode of production (Harvey, 1996), and neoliberalism, as capitalism's contemporary phase, is inclined to crisis due to capital over-accumulation. As Cocharne and Ward state (2012: 7), the 'significance of neoliberalism [. . .] cannot be understood as a sort of global dust cloud just waiting to settle somewhere to be realised in some fixed and more or less perfect form'. In the EU, due to uneven geographical development, the crisis has mostly affected the peripheral regions of Southern Europe, and especially Greece (Hadjimichalis, 2010). Arguably, the seeds of the economic crisis lie within the 2007–2009 period through speculative mortgage lending by US financial institutions and the trading of derivatives by international banks, together with the precarious ways peripheral countries were integrated into the Eurozone (Lapavitsas et al., 2010). By the end of 2009, a dramatic decline in public revenues pushed Greece, together with other southern European countries, to borrow money from international institutions. In May 2010, the newly elected social democrat government decided that a rescue from the fiscal crisis would be possible via a bailout agreement under a borrowing programme from the IMF and the EU (Kouvelakis, 2011; Maloutas, 2009). The government's reasoning for this agreement was based on a belief that the country had been financially mismanaged for years (Souliotis, 2013), disregarding any possible understanding of the crisis as a result of the neoliberal policies promoted by institutions such as the EU. The role of northern European surpluses and their close connection with southern European deficits (Lapavitsas et al., 2010) was also ignored.

   Both the first (2010) and the second (2012) Memorandum of Agreement with what is commonly known as the Troika (the EU, IMF and ECB creditors) proposed a package of significant budget cuts and new taxation. In other words, an austerity model, understood as the means by which costs of macroeconomic mismanagement and financial speculation are visited upon the dispossessed and the disempowered (Peck, 2012). In Greece, the austerity policies imposed through the Memorandum sought a way to save the banking sector whilst transferring the costs to middle and lower classes (Maloutas et al., 2012). The neoliberal stimulus bowed to the creative destruction of capitalism, with public land and assets portrayed as prime attractions for foreign investment, labour costs expected to diminish, environmental legislation easily bypassed and urban infrastructures – such as electricity and water – perceived as a means of extracting further revenues from citizens.

Within this framework, another means of surveillance and social control was established as the politics of fear through economic tools – such as taxes – was enforced against anti-austerity disobedience. Such imposed measures brought one of the most drastic drops in living standards in post-war Europe. Workers and pensioners lost around a third of their incomes (Kouvelakis, 2011) whilst schools were left without textbooks and the quality of public services deteriorated sharply. In Athens' most deprived neighbourhoods, school children were collapsing due to hunger while hospital patients were advised to bring their own sheets and medicine. By 2014 the official unemployment rate was estimated to be at least 27 per cent (Eurostat, 2014), with data indicating that this rate nearly doubles for young people. Suicide rates in Greece, among the lowest in Europe prior to the crisis, were estimated to have doubled during the 2011 to 2013 period.

The focus of the Memorandum on wage decreases, increased taxation and the introduction of new taxes on electricity bills can be understood in tandem with the tendency of capitalism to accumulation by dispossession (Harvey, 2007). The end result of such neoliberal policies and practices is the inability of the household to pay the bills and the appropriation of the households' material and immaterial property. Indignation and frustration against austerity transformed into a political opportunity for collective action (Tarrow, 1994). Waves of demonstrations, general strikes and social mobilisations emerged to oppose the neoliberal doctrine. In Athens, in the summer of 2011, a group of citizens formed a movement called the Aganaktismenoi of the Syntagma Square, echoing similar social mobilisations worldwide such as the Indignants of Tahrir Square in Egypt and the Indignados of the Puerta del Sol in Madrid (Abellan et al., 2012; Leontidou, 2012).

Those who gathered in Syntagma Square from May until July 2011 were a heterogeneous group, establishing two distinctive forms of protest and occupation. Protesters in the upper square would sing football match-like slogans against the government, whilst those on the lower square actively engaged in discussion and political thinking through debates on questions of direct democracy. By establishing a daily assembly where everyone could participate, the movement of the lower square opened discussion around issues such as living conditions, employment rights, constitutional and economic principles, public debt and the legitimacy of the austerity policies. This assembly organised demonstrations and manifestations that challenged austerity and the credibility of the creditors. By August 2011 the assembly of Syntagma Square was violently suppressed by the police, as activists occupying the public space were removed through arrests and the widespread use of tear gas. However, the spirit and ideas that characterised the assembly at the lower end of the square were transferred to the neighbourhoods of Athens, where activists formed assemblies at the neighbourhood level or enhanced pre-existing local assemblies. This enriched the general arguments calling for direct democracy and challenging austerity measures and extensive taxation. It is in these local assemblies where the taxation on electricity, the Charatsi, was initially discussed, and the space from where the new protest culture against the Charatsi tax emerged.

The Charatsi: What Is It?

The Public Power Corporation of Greece (or DEI Company), the state entity in charge of electricity provision, operates through mixed ownership. With 51.12 per cent of its stocks directly in the hands of the Greek state, it recognises electricity as a public good (DEI, 2014b). According to the company's website, the DEI Company 'was established to operate in the interests of the public' (DEI, 2014a). The aim of the company is to implement a national energy policy, which would enable every Greek citizen to make use of electrical power 'at the lowest possible price'.

However, in September 2011, due to 'urgent national reasons that forge the need to minimise the public deficit' (Governmental Gazette, Law 4021/2011), the Greek government approved ad hoc taxation that would help the state to gather additional revenue aimed at paying off its international creditors. One of these measures was directly linked to electricity consumption, resulting in a new tax on household electricity bills. Named by the government as the Emergent Special Tax on Electrical Power Supplied to Built Spaces, it became popularly known as the Charatsi on power supply. It was calculated through a series of coefficients relating to a property's area, land zone, size and age. Critically, the landowner's income was not taken under consideration in the calculation – for example, a deprived household in a middle-class neighbourhood living in an inherited house with a large area has to pay a large tax. Exceptions included buildings that belong to the state, the church, embassies, buildings that are used for educational, cultural, religious and public benefit purposes, stadiums and sporting installations. Additionally, listed buildings, archaeological sites, as well as land plots used for industrial and agricultural purposes, were declared to be exempt. The law included limited criteria for tax exemptions or reduced taxation for specific vulnerable social groups, including exemptions for those on limited incomes, handicapped citizens and households with over four children. According to Law 4021/2011 (Greece, Governmental Gazette, Law 4021, 2011), the tax would be collected by the DEI Company in two instalments. The first one was arranged for the end of October 2011 and the second by the end of January 2012. The DEI Company announced that households who refuse to pay the tax would be treated in a punitive way, facing disconnection from the electricity grid. Reconnection would be established once the Charatsi tax bill was settled.

For the first time in the history of the modern Greek state a spatially orientated tax via electricity consumption was imposed. In an indirect way, the tax penalised homeownership; the bigger the size of the property, the greater the amount of tax imposed. If households were unable to pay the added taxation, their ownership of the property would be jeopardised. The Charatsi's impact needs to be examined in the context of homeownership in Greece. Homeownership is not related to an upper-middle class status but rather to the popularisation of forms of financial security against the economic and political instability characteristic of the post-war period. The advent of urbanisation models such as the *antiparochi*, a construction system which replaced low-story housing stock with a high rise urban model based on co-ownership (Maloutas et al., 2012), enabled widespread access to homeownership whilst avoiding large-scale and state-led social housing programmes. For the urbanizing lower- and middle-class households, homeownership in *antiparochi* apartments functioned as a safety net against the political instability, low wages and precarity of the labour market. As a result, homeownership is not necessarily a marker of wealth concentration, but rather a result of historic urban processes towards the development of safety nets. The imposition of the Charatsi, and its particular mode of implementation, signalled an attempt to reshape the country's property map, as households unable to pay the added tax faced foreclosure. Here accumulation by dispossession, as outlined by Harvey (2007), takes place, as property ownership is transferred to banks and realtors. The future of the property market and the luck of the bankrupt households were determined by the futures of neoliberalism and its impetus to form new markets and impose new doctrines.

A brief analysis of why this new tax became popularly known as the Charatsi is telling of how it is perceived by the public. The etymological root of the word 'Charatsi' can be traced to Arabic. According to Islamic Law, the ḵarāj (خراج) is the 'land tax'. During the Ottoman Empire, non-Muslim residents were required to pay an annual per capita tax to the Sultan. Usually tax collectors would request a larger amount of money in order to personally benefit from this collection regime. The word 'Charatsi' was revived by the Greek public, foregrounding the injustice of new taxations resulting from the emergency measures associated with the negotiations between the state and the IMF, the EU and the ECB – or the Troika. For the public, the concept of Charatsi reflects a historical reading of an illegal, unjust theft by outside conquerors. This new tax on the power supply was understood by society to be a type of Charatsi, similar to the one imposed during the Ottoman Empire. The added tax on electricity consumption has put an extra burden on working-, lower-middle- and middle-class homeowners, segments of society who already have suffered major decreases in their net incomes and consumption standards. This is particularly the case for the most impoverished households, often living in dwellings characterised by (energy) poor conditions (no building insulation, no heating and so forth). In some cases reported by the media, this form of electricity rationing led to poor households relying on fire stoves for heating purposes, which have on occasion proved lethal (TVXS, 2014). The burning of a variety of materials for heating purposes – such as processed wood containing plastic and other chemical compounds – an increasingly common practice amongst poorer households, contributed to a deterioration in air quality in large cities. In protest at these conditions, many people became active in the local anti-Charatsi assemblies, feeling that their right to electricity, housing and quality of life was directly at stake.

In our methodology we embraced an ethnographic approach alongside an in-depth review of the publicly available literature related to the crisis and local movements in Athens. The ethnographic component included participant observation and active engagement in two local anti-Charatsi assemblies. Ethnographic methods, in capturing people's social meanings and activities (Brewer, 2000), are appropriate for generating direct encounters with agents and their social conduct (Willis & Trondman, 2000). In our research, we wanted to achieve a direct and continuous interaction with activists. Those participating in assemblies were informed about the purpose of our academic research and the expected outcomes in the form of academic papers unpacking the emergent protest culture. We rejected the idea of conducting interviews with other fellow activists for ethical reasons, as we were constantly interacting with them both via informal conversations and through participation in the assemblies, often on a daily basis. Prioritising participant observation over interviews helped to avoid the creation of feelings of mistrust or misconceptions resulting from the inquisitive nature of academic research. Often members of the general public participating in local assemblies feel frustrated and exploited by academics who conduct research in the name of academic activism yet barely interact with the communities and values at stake and may misinterpret activists. As such, we preferred to write memos and detailed notes after every assembly and protest, exchange thoughts and develop our ideas within the assemblies. We enriched the ethnographic approach through an ongoing examination of newspapers, the evolving legal framework, Internet blogs and web articles related to the crisis, mobilisations and especially the anti-Charatsi protests.

The Anti-Charatsi Mobilisations and Protests

The anti-Charatsi mobilisations introduced new protest cultures that bear little relation to the way mobilisations had previously been established in Greece. Until the 1990s, due to clientelism and weak state traditions, local movements were easily manipulated by the state (Leontidou, 2010). However, since the late 1990s, local social movements emerged in response to the neoliberalisation of space and associated deterioration of living conditions in big cities such as Athens (Kavoulakos, 2013). Such a shift in their rationale liberated them from state control. As discussed by Arampatzi and Nicholls (2012), by the beginning of 2000s, the Greek state appeared to possess diminished capacities to co-opt the anti-neoliberal resistance that emerged. Since then, local movements have become spaces of interaction and networking, forging a degree of trust and solidarity amongst members.

A decade later, the Memorandum agreement signed with the Troika in 2010 and 2012 and the imposition of heavy taxation may be interpreted as a further rupture in the social contract established since the post-war period. Newly formed civil society groups and networks are developing a form of cultural politics examining the world with a highly critical gaze (Afouxenidis, 2015). The anti-Charatsi mobilisations sprung out at a moment when the state could hardly control the anger of middle and working classes who were experiencing a loss in basic forms of service provision previously established by the welfare estate – where a threat to the universality of electricity services drove them towards social protest and resistance.

The way that the various forms of mobilisation by different sectors of Greek society have interacted and intersected is indicative of a type of openness within the new protest cultures that have emerged after the crisis. After the announcement of the new tax on the consumption of electricity, civil society reacted in several ways. Citizens, either individually or as members of local anti-Charatsi assemblies, filed appeals in the civic courts claiming the inability or disavowing the need to pay. These appeals were often prepared by lawyers' associations claiming the illegality of the new tax. The Athens' Bar Association, the association of lawyers who practice in Athens, reacted against the Charatsi by publicly denouncing it in the media and officially to the Supreme Court of Council of the State of Greece (the Supreme Court on Civil issues, or *Symvoulio tis Epikrateias*). Lawyers cooperated with municipalities, as in the case of the municipality of Kallithea and Nea Ionia, two cities within the Athens' conurbation, guiding members of the public on how to pay electricity bills excluding the amount that corresponds to the Charatsi. In parallel, legal verdicts argued over the illegality of the new tax; a third tier of civic courts judgments approved cases against the state, ruling in favour of members of the public who could not afford to pay the tax and others whose Charatsi was miscalculated (TVXS, 2012a). By March 2012 the Council of State of Greece adjudicated on the legality of the Charatsi, but declared the illegality of disconnecting those unable to afford payment of the full amount of the electricity bills (Bulletin of Tax Legislation, 2012). In February 2014, under strong grassroots pressure and in the context of negative reactions from the broader society, the Supreme Civil and Criminal Court of Greece (*Areios Pagos*) declared the illegality of the Charatsi and announced its opposition to it at the European Convention on Human Rights.

The reaction of both civil society and local governments against the Charatsi tax has been widespread. In many cities both elected officials as well as staff members of the local government, and in some cases mayors, participated in symbolic occupations of public buildings and other protest activities directed against the local electricity supply offices of the DEI Company (TVXS, 2011). They collaborated with advocacy groups to provide residents with arguments against the Charatsi tax payments (Karatziou & Polychroniadis, 2011). In some cases councillors and mayors who actively engaged in the anti-Charatsi protests were characterised as 'irresponsible' and had to face a process of 'attribution of disciplinary responsibilities' (Vourgana, 2011). The main political parties opposed to the Charatsi were the left-wing political coalition of Syriza and the Greek Communist Party. Representatives of these parties persistently denounced the Charatsi and openly supported public protests against it. Under pressure to respond to both protesters and the public, the right-wing party New Democracy promised during the 2012 election campaign to abolish the Charasti, alongside other changes in the taxation system. However, these promises were not implemented after the party's victory in these national elections. Other sectors of society, such as commercial associations and civil society organisations, publicly expressed opposition to the Charatsi. The Charatsi was seen by commercial associations as a direct and indirect burden on economic activity across the country, imposing additional outlays on entrepreneurs already dealing with an increase in costs and a drop in the purchasing power of the population. In 2012 the National Confederation of Hellenic Commerce declared that many households and companies could not afford to pay their electricity bills, requesting an immediate withdrawal and decoupling of the Charatsi and the electricity bill (National Confederation of Hellenic Commerce, 2012).

The trade union of the DEI Company, which represents the interests of workers within the electricity sector, also expressed strong views against the Charatsi. The union organised mobilisations against the Charatsi by calling on the public not to pay the tax, and put pressure on the government to withdraw the measure. In 2011 the unionists performed two symbolic actions against the Charatsi, starting with a symbolic occupation of the Ministry of Health and Welfare (which had an outstanding unpaid electricity bill of €3.8 million) whilst also disconnecting the building's power supply for four hours. During this event, the union claimed that the DEI Company's main debtor was the Greek state itself, which owed over €140 million in unpaid bills (Eleftherotypia, 2011). The union's second big mobilisation was at the DEI Company's operations centre for electricity supply, where the archives of unpaid Charatsi bills were kept. Unionists occupied the building for over three days, an industrial action that ended with the arrest of the president as well as other members of the union (Capital, 2011). After this event the president of the union released the names of institutions and wealthy individuals who did not pay the Charatsi to the media.

In 2011 members of the public formed the 'I do not pay' movement, developing a specific mobilisation against the Charatsi payments and other cost increases in public services. The movement called on members of the public to refuse to pay the increase in the tolls on national roads, and subsequently collaborated with trade unions, lawyers' associations and local movements in order to raise awareness and increase the battle against Charatsi payments (Figures 10.1 and 10.2).

Furthermore, bottom-up initiatives mobilising against the Charatsi payments emerged at the local level. Local assemblies would take place once a week at designated places in the neighbourhood (such as squats and open public spaces). These were not hierarchical but followed horizontal structures where decision-making was reached through consensus. The issues raised prioritised a variety of anti-Charatsi practices: appeals against the courts, power reconnections and forms of avoiding Charatsi payments. The latter included limiting the bill payment to the amount corresponding to electricity consumption – in this way disregarding the amount related to the Charatsi – to be paid at banks' cash points (where the electricity user decides on the total amount to be paid) or at specific DEI Company payment offices where the directors, in solidarity with the movement, were willing to accept partial payments. Participants in the assemblies communicated with each other via ICT technologies that would notify everyone in case of an important event or moment, such as a power reconnection in the neighbourhood. Some assemblies put pressure on the local government of Athens requesting local officers to choose sides between what they deem to be the wishes of local inhabitants or those of the Troika.

Figure 10.1 Local assemblies mobilise in Athens
Source: the authors

Figure 10.2 Local assemblies mobilise in Athens
Source: the authors

The broader neighbourhood received information from assembly members through visual means, including posters, leaflets and flyers handed out by activists. Slogans circulating via posters and leaflets include phrases such as:

Electricity is a common good;

No household should be left out without electricity;

Break down the fear, join us;

Let's resist their blackmails, abolish the Charatsi and reconnect electricity;

Bankrupts of the world, unite!

Such information was circulated in places frequently used by the inhabitants, such as the weekly open market, the local supermarket or in front of the entrance to the local offices of the DEI Company where members of the public pay their bills. Leaflets also contained legal details and information, inviting members of the public to gatherings and local demonstrations.

Handouts included emergency contact details for the assembly, so that those in need could get in touch and, for example, ask for support towards a reconnection. Most of the assemblies quickly acquired the technical equipment required to carry out reconnections. Since power supply reconnections by individuals are considered to be 'illegal', assemblies undertook full responsibility for these actions; where a household was disconnected due to unpaid bills, local assemblies, once informed, would reconnect the power supply. In addition, YouTube videos uploaded by protesters provided advice on the technical aspects of power reconnections. Such initiatives gained legitimisation in the local communities of Athens. For example, between 2011 and 2014 the anti-Charatsi Assembly of Exarchia, a neighbourhood in the centre of Athens, undertook 53 power reconnections (Assembly of Exarchia, 2014). Moreover, the daily flow of calls to the assemblies related not only to electricity but also to issues around food, clothing, mortgages and taxation. In Athens, representatives of neighbourhood assemblies gathered once a month, sharing experiences and transferring knowledge. Assembly meetings, often held in plazas, parks or other public grounds, repositioned public space as a site of social and political struggles around resource use (see Mitchell, 2003).

Government Reactions

By mid-2012, over 500,000 Greeks had not paid the Charatsi. In June 2012 the unpaid amount was calculated to be €1.1 billion (TVXS, 2012b). This pressured the central government to change the Charatsi's payment arrangements so that it could be paid in five instalments instead of the two initially foreseen (Greece. Governmental Newsletter, 2011). This was celebrated in the local assemblies, and activists considered the move as both a victory and a sign of success in tearing down the general sense of 'fear' imposed in an era of austerity and crisis.

However, limiting the sense of victory, the DEI Company changed its bill collection mechanism to ensure that the paid section of the bill would always include the Charatsi payment, even in the case of partial payments. This meant that the new bill payment system would register the Charatsi payment first, followed by a – complete or incomplete – electricity payment, deterring users who had previously limited their payment to the costs associated with electricity consumption but refused to pay the added tax. Additional indirect taxes were introduced to the electricity bills, this time related to the payment of EU fines which the DEI Company incurred resulting from environmental regulations and economic restrictions against monopolies. By the end of 2012, in a move to increase control and suppress the anti-Charatsi mobilisations, the collection of the Charatsi was transferred from the DEI Company to the Tax Revenue Office. Households which had not paid the Charatsi were forced to settle accounts in order to avoid wider problems with tax eligibility. It was determined that those households which, after 2014, still owed debts to the state, including unpaid electricity and Charatsi bills as well as other taxes, could face government seizure of goods for the purpose of payment. In the case of debts of over €5,000, imprisonment and property or salary seizure could apply (TVXS, 2012b). Law 3869/2010 (Greece. Governmental Gazette, Law 3869, 2010), forbidding the auction of the primary residence of heavily indebted households, was deemed void from December 2013 onwards. Henceforth, people who could not pay off mortgages and debts to the state lost their properties. By May 2014, 10,000 foreclosures were ordered against indebted households. In cases where bank savings and salary contributions were not enough to complete the payment, the state had full jurisdiction to evict families. Yet, in spite of these threats, Athens' grassroots movements against electricity taxes have continued to advocate against payment. According to activists, further imposition of Charatsi taxes on other basic services and public infrastructures (including water, housing and sanitation) are to be expected (Marvas, 2014). However, following the 2014 election of the Syriza government, the Minister of Development suspended foreclosures of primary residences (or what in Greece is known as the first house, declared to the state as the primary residence) due to unpaid Charatsi and bills (TVXS, 2015).

From the Anti-Charatsi Mobilisation to Solidarity Cultures

After the collection of Charatsi was transferred to the Tax Revenue Office, mobilisations against it slowed down. This meant that Charatsi payments became unavoidable, as all citizens have to deal with it through their income declaration. In the summer of 2014 the Charatsi, initially imposed as a temporary tax, was made permanent and turned into a direct tax: the Enfia, otherwise known as the Consolidated Tax on Property Ownership. The Enfia operates as an additional property tax, independent of traditional property taxation, via income declaration. This new form of Charatsi is no longer related to electricity services and power consumption, but, via the Enfia, is an additional tax imposed on property. The Charatsi, a temporary government measure to raise revenue for the payment of public debt, became a permanent presence within a three-year period.

Beyond austerity and heavy taxation – creating fear and despair in much of the population – the local anti-Charatsi mobilisations opened spaces of social interaction and support across and beyond intervening in electricity infrastructures. The dynamic nature of the assemblies transformed the anti-Charatsi discussion into a broader anti-austerity mobilisation. Most local assemblies engaged in initiatives aimed at responding to the government's austerity efforts, including soup kitchens, food collection and distribution, clothes exchange and the establishment of medical centres and mini-markets. In many neighbourhoods in Athens, after-school lessons for children were established, serving those who need additional support to improve performance and whose families could not afford private tutoring. Alternative economic and commercial models have been promoted based on direct producer-consumer transactions, such as market systems where producers sell directly to consumers without the intervention of intermediaries. These novel forms of social organisation have led to the creation of social bonds and intensive networking at the local level, where ideas and practices of solidarity have become part of the everyday battle against austerity. Solidarity is put forward both as a way of collaborating and a collective reaction against the crisis. In contrast to philanthropy, where the middle classes show their sympathy with vulnerable groups, solidarity structures put forward the idea of an alternative parallel organisation. In Athens, solidarity involved everyday embodied practices such as local assemblies and daily initiatives dealing with food distribution or cancelation of auctions, while constituting a shared experience and becoming a driving force of motivation, mobilisation and cooperation among participants (Arampatzi, 2015). The emerging social structures of solidarity alleviate socio-economic burdens and protect the public against punitive forms embedded in capital accumulation, particularly for the underprivileged. Those affected by the crisis thus build alternative networks that create spaces of hope. Through alternatives to austerity, solidarity becomes a generative force as it enables the creation of new spaces for social empowerment (Arampatzi, 2015) that may actually 'crack capitalism' (Holloway, 2010).

Conclusion: Some Reflections on the Charatsi

This chapter has argued that, in light of an understanding of electricity provision as a public good and of the public nature of Greece's primary electricity supplier, the Charatsi tax translates into an attack on public and common goods. Such attacks are significant, illustrating the extent to which public goods such as electricity or basic goods such as housing are at stake under austerity regimes. However, this – arguably illegal – tax imposed on electricity consumption was met by Greek society with an outcry. The Charatsi triggered mobilisations by a broad variety of institutional and non-institutional actors, from individuals to lawyers' associations, trade unions, municipalities, bottom-up initiatives and local assemblies.

The Charatsi illustrates how new forms of taxation introduced within the private sphere (the house, via homeownership) via basic services and infrastructures of provision – such as electricity – can soon become a public issue galvanising society's efforts in a battle against austerity and further taxation. Informal reconnections of power supply carried out by local assemblies obtained legitimacy through popular support, as many households were reluctant to pay the tax. Households under the threat of disconnection relied on the help of local assemblies and lawyers' associations for appeals against the courts. The resulting social networks provided novel forms of experimentation with societal models based on solidarity and support.

In Athens, the anti-Charatsi mobilisations supported the development of new protest cultures. Since the 2000s, social movements in Greece have developed a unique character, differentiating them from the strong unionism and political party focus prevalent in the past (Kavoulakos, 2013). We suggest that the anti-Charatsi protests further expand this trend. With the anti-Charatsi movement, actors with significantly different standpoints, such as municipalities, courts and local assemblies, established collaborative agendas against austerity whilst developing a new portfolio of social practices. These include intense networking and the establishment of solidarity structures, all of which are used towards enhancing the effectiveness of the joint social agendas at play.

Inevitably, the austerity government of a state in crisis has responded through strengthening surveillance techniques aimed at controlling social movements (Souliotis & Kandylis, 2013). In the anti-Charatsi case study, the state reallocated the amounts repaid on electricity bills favouring the tax charge over the service supplied (power consumption), as a mechanism to suppress the mobilisation. In some cases, the local offices of the DEI Company were equipped with private security guards in order to prevent harassment and disruption from people's mobilisations. Importantly, the transfer of Charatsi collection to the Tax Revenue Office established at the end of 2012 paved the way for its 2014 incorporation as an additional property tax in its own right. We argue that this political tactic has been manifold. It introduced a mode of attack to the savings and income of the majority of households, in order to secure revenues for the debt payoff. Moreover, it served as an additional means of dispossession, as the properties of indebted households were transferred to real estate agencies, banks and the state. In the end, the governmental manoeuvres for the collection of the tax led to the effective control and elimination of the anti-Charatsi mobilisations.

However, the local assemblies that engaged in the anti-Charatsi movement enriched both protest discourse and practices, as they developed solidarity networks and structures against the crisis and offered alternative economic and social systems. By establishing solidarity structures that function in parallel to those established by the state, a form of collective thinking and interaction is introduced, effectively operating as a 'termite' in the institutional and material formation of capitalism (Harvey, 2012). An understanding of electricity as a basic affordable and universal service, as much as the social medical centres, local markets (without intermediaries) and soup kitchens established by the assemblies, challenges the rules of a market economy and the pervasive idea of commodification. The model generated within, and advocated by, the assemblages puts forward alternative social and economic systems based on the principles of solidarity and cooperation, in this way providing alternatives to the dominant neoliberal perspective. As such networks gain power, new opportunities for the creation of spaces of hope emerge, replacing the austerity of despair.

References

Abellan, J., J. Sequera and M. Janoschka, 2012. 'Occupying the #Hotel Madrid: A Laboratory for Urban Resistance'. *Social Movement Studies* 11(3–4): 320–36.

Afouxenidis, A., 2015. Neoliberalism and Democracy. In: J.K. Dubrow (ed.), *Political Inequality in the Age of Democracy: Cross National Perspectives*. London: Routledge, pp. 40–8.

Arampatzi, A., 2015. *Resisting Austerity: The Spatial Politics of Solidarity and Struggle in Athens, Greece.* Department of Geography, Leeds University. PhD.

Arampatzi, A. and W. Nicholls, 2012. 'The Urban Roots of Anti-neoliberal Social Movements: The Case of Athens, Greece'. *Environment and Planning A* 44(11): 2591–610.

Assembly of Exarchia, 2014. Minutes of the 18 June 2014 DIKAEX Assembly, Athens. [Circulated to Assembly Members Via e-mail List].

Bouzarovski, S., 2014. 'Energy Poverty in the European Union: Landscapes of Vulnerability'. *Wiley Interdisciplinary Reviews: Energy and Environment* 3(3): 276–89.

Brewer, J., 2000. *Ethnography*. Buckingham: Open University Press.

Capital, 2011. Fotopoulos Was Arrested and Sent to the Advocate [in Greek]. [Online]. 24 November. Available from: http://www.capital.gr/news.asp?id=1338899 [Accessed: 25 November 2011].

Cocharne, A. and K. Ward, 2012. 'Researching the Geographies of Policy Mo-bility: Confronting the Methodological Challenges'. *Environment and Planning A* 44(1): 5–12.

DEI, S.A., 2014a. Public Power Corporation S.A. Today [in Greek]. [Online]. Available from: http://www.dei.com.gr/en/i-dei/i-etairia/omilos-dei-ae/dei-ae [Accessed: 15 July 2014].

DEI, S.A., 2014b. DEI Shared Capital [in Greek]. [Online]. Available from: http://www.dei.gr/el/xrimatistiriaka-stoixeia/metoxiki-sunthesi [Accessed: 4 May 2014].

Eleftherotypia, 2011. GENOP-DEI Disconnected the Power-Supply [in Greek]. [Online]. 22 November. Available from: http://www.enet.gr/?i=news.el.ellada&id=326592 [Accessed: 23 November 2011].

Eurostat, 2014. Eurostat Unemployment Statistics. [Online]. Available from: http://epp.eurostat.ec.europa.eu/statistics_explained/index.php/Unemployment_statistics#Recent_developments_in_unemployment_at_a_European_and_Member_State_level. [Accessed: 13 September 2014].

Greece. Bulletin of Tax Legislation, 2012. The Ratification of the Disconnection of Power Supply is Unconstitutional [in Greek]. [Online]. Available from: http://dfn.gr/files/pdf/neo.antisint.eetide.4.3.12.pdf [Accessed: 18 April 2012].

Greece. Governmental Gazette Law 3869/2010, 2010. Setting the Debts of Indebted Individuals and Other Provisions [in Greek]. [Online]. Available from: http://www.efpolis.gr/el/library2.html?func=fileinfo&id=226 [Accessed: 6 May 2013].

Greece. Governmental Gazette Law 4021/2011, 2011. Enforced Measures for the Control and Purge of Institutions, Regulations on Financial Issues and Other Measures [in Greek]. Athens the Statutory Office. [Online]. Available from: http://www.hellenicparliament.gr/Nomothetiko-Ergo/Anazitisi-Nomothetikou-Ergou?law_id=96a46802-d7ce-4477-92cf-f845b79275d6 [Accessed: 3 March 2012]

Greece. Governmental Newsletter, 2011. Further Instructions on the Implementation of L. 4021/2011, article 52, 1/12/2011 [in Greek]. [Online]. Available from: http://static.old.

diavgeia.gov.gr/doc/%CE%9244%CE%95%CE%97-%CE%A008 [Accessed: 15 March 2012].

Hadjimichalis, C., 2010. 'Uneven Geographical Development and Socio-spatial Justice and Solidarity: European Regions After the 2009 Financial Crisis'. *European Urban and Regional Studies* 18(3): 254–74.

Harvey, D., 1996. *Justice, Nature and the Geography of Difference*. Oxford: Blackwell Publishing.

Harvey, D., 2007. 'Neoliberalism as Creative Destruction'. *The Annals of the American Academy* 610(1): 22–45.

Harvey, D., 2012. *Rebel Cities: From the Right to the City to the Urban Revolution*. London: Verso.

Holloway, J., 2010. *Crack Capitalism*. New York: Pluto Press.

Karatziou, D. & M. Polychroniadis, 2011. Troika Expresses Worries Over the Charatsi [in Greek]. [Online]. *Elefterotypia*. 20 November. Available from: http://www.enet.gr/?i=news.el.ellada&id=327320 [Accessed: 15 December 2011].

Kavoulakos, K., 2013. Movements and Public Spaces in Athens: Spaces of Freedom, Spaces of Democracy, Spaces of Dominance [in Greek]. In: T. Maloutas, G. Kandylis, M. Petrou and N. Souliotis (eds.), *The City Centre of Athens as a Political Stake*. Athens: EKKE, pp. 237–56.

Kouvelakis, S., 2011. 'The Greek kaldron'. *New Left Review* 72(1): 17–32.

Lapavitsas, C., A. Kaltenbrunner, G. Labrinidis, D. Lindo, J. Meadway, J. Michell, J.P. Painceira, E. Pires, J. Powell, A. Stenfors, and N. Teles, 2010. The Eurozone between Austerity and Default. Research on Money and Finance (RMF) occasional Report. September 2010. [Online]. Available from: http://www.researchonmoneyandfinance.org/. [Accessed: 4 February 2016].

Leontidou, L., 2010. 'Urban Social Movements in "Weak" Civil Societies: The Right to the City and Cosmopolitan Activism in Southern Europe'. *Urban Studies* 47(6): 1179–203.

Leontidou, L., 2012. 'Athens in the Mediterranean "Movement of the Piazzas"; Spontaneity in Material and Visual Public Spaces'. *City* 16(3): 299–312.

Maloutas, T. (2009). 'Urban Outcasts: A Contextualized Outlook on Advanced Marginality'. *International Journal of Urban and Regional Research* 33(3): 828–34.

Maloutas, T., V. Arapoglou, G. Kandylis and J. Sayas, 2012. Social Polarisation and De-segregation in Athens. In: T. Maloutas & K. Fujita (eds.), *Residential Segregation in Comparative Perspective: Making Sense of Contextual Diversity*. Surrey, Burlington: Ashgate, pp. 257–84.

Marvas, D., 2014. 'Enfia, Ignorance, Hypocricy or Something Different?'. *TVXS*. [Online]. 28 August. Available from: http://tvxs.gr/news/egrapsan-eipan/enfia-agnoia-ypokrisia-i-kati-allo [Accessed: 3 September 2014].

McFarlane, C. and J. Rutherford, 2008. 'Political Infrastructures: Governing and Experiencing the Fabric of the City'. *International Journal of Urban and Regional Research* 32(2): 363–74.

Mitchell, D., 2003. *The Right to the City: Social Justice and the Fight for Public Space*. London: The Guilford Press.

National Confederation of Hellenic Commerce, 2012. Disconnect the Emergent Special Tax on Electrical Power Supplied Built Spaces from the Bills of Power Supply [in Greek]. 6 March. [Online]. Available from: http://www.esee.gr/page.asp?id=3946 [Accessed: 24 March 2013].

Peck, J., 2012. 'Austerity Urbanism: American Cities Under Extreme Economy'. *City* 16(6): 626–55.

212 Georgia Alexandri and Venetia Chatzi

Peck, J., N. Theodore and N. Brenner, 2009. 'Neoliberal Urbanism: Models, Moments, Mutations'. *SAIS Review of International Affairs* 29(1): 49–66.
Souliotis, N., 2013. Athens and the Politics of the Sovereign Debt Crisis. In: K. Fujita (ed.), *Cities and Crisis: New Critical Urban Theory*. Washington: Sage, pp. 237–69.
Souliotis, N. and G. Kandylis, 2013. Athens and the Politics of the Sovereign Debt Crisis. In: *Interrogating Urban Crisis: Governance, Contestation and Critique*. De Montfort, Leicester, Monday 9 to Wednesday 11 September.
Tarrow, S., 1994. *Power in Movement: Social Movements, Collective Action and Politics*. New York: Cambridge University Press.
TVXS, 2011. The Municipalitices Don't Pay the Charatsi [in Geek]. [Online]. 12 November. Available from: http://tvxs.gr/news/ellada/den-plironoyn-den-plironoyn-oi-dimoi-xaratsi-tis-dei [Accessed: 16 March 2012].
TVXS, 2012a. Illegal the Collecting of Charatsi Via the DEI [in Geek]. [Online]. 4 December. Available from: http://tvxs.gr/news/ellada/paranomi-i-eispraksi-toy-xaratsioy-meso-tis-dei [Accessed: 10 February 2013].
TVXS, 2012b. Seizure of Bank Accounts for Debts More than 4.000€ [in Greek]. [Online]. 4 March. Available from: http://tvxs.gr/news/ellada/katasxesi-trapezikon-logariasmon-gia-xrei-stin-eforia-ano-ton-4000-eyro [Accessed: 2 September 2013].
TVXS, 2014. Tetraplegic Woman Died When DEI Disconnected the Electricity Power [in Greek]. [Online]. 23 June. Available from: http://tvxs.gr/news/ellada/tetrapligiki-gynaika-pethane-molis-i-dei-tis-ekopse-reyma [Accessed: 25 June 2014].
TVXS, 2015. Stathakis: Direct Regulations of the Households' Red Loans [in Greek]. [Online]. 12 February. Available from: http://tvxs.gr/news/ellada/stathakis-amesi-rythmisi-ton-kokkinon-daneion-gia-ta-noikokyria [Accessed: 12 February 2015].
Vourgana, M., 2011. Favorable Settings for the Emergent Special Tax on Electrical Power Supplied Built Spaces [in Greek]. *Imerisia*. [Online]. 18 November. Available from: http://www.imerisia.gr/article.asp?catid=26516&subid=2&pubid=112653569 [Accessed: 15 December 2011].
Willis, P. and M. Trondman, 2000. 'Manifesto for Ethnography'. *Ethnography* 1(1): 5–16.

Index